Paul J. Mlynarczyk

# The Dynamic Glass Transition Temperature in Polymeric Liquids

AF138185

Paul J. Mlynarczyk

# The Dynamic Glass Transition Temperature in Polymeric Liquids

## Characterization and Measurement

LAP LAMBERT Academic Publishing

**Impressum / Imprint**
Bibliografische Information der Deutschen Nationalbibliothek: Die Deutsche Nationalbibliothek verzeichnet diese Publikation in der Deutschen Nationalbibliografie; detaillierte bibliografische Daten sind im Internet über http://dnb.d-nb.de abrufbar.
Alle in diesem Buch genannten Marken und Produktnamen unterliegen warenzeichen-, marken- oder patentrechtlichem Schutz bzw. sind Warenzeichen oder eingetragene Warenzeichen der jeweiligen Inhaber. Die Wiedergabe von Marken, Produktnamen, Gebrauchsnamen, Handelsnamen, Warenbezeichnungen u.s.w. in diesem Werk berechtigt auch ohne besondere Kennzeichnung nicht zu der Annahme, dass solche Namen im Sinne der Warenzeichen- und Markenschutzgesetzgebung als frei zu betrachten wären und daher von jedermann benutzt werden dürften.

Bibliographic information published by the Deutsche Nationalbibliothek: The Deutsche Nationalbibliothek lists this publication in the Deutsche Nationalbibliografie; detailed bibliographic data are available in the Internet at http://dnb.d-nb.de.
Any brand names and product names mentioned in this book are subject to trademark, brand or patent protection and are trademarks or registered trademarks of their respective holders. The use of brand names, product names, common names, trade names, product descriptions etc. even without a particular marking in this works is in no way to be construed to mean that such names may be regarded as unrestricted in respect of trademark and brand protection legislation and could thus be used by anyone.

Coverbild / Cover image: www.ingimage.com

Verlag / Publisher:
LAP LAMBERT Academic Publishing
ist ein Imprint der / is a trademark of
OmniScriptum GmbH & Co. KG
Heinrich-Böcking-Str. 6-8, 66121 Saarbrücken, Deutschland / Germany
Email: info@lap-publishing.com

Herstellung: siehe letzte Seite /
Printed at: see last page
**ISBN: 978-3-659-55506-0**

Zugl. / Approved by: Manhattan, Kansas State University, 2014

Copyright © 2014 OmniScriptum GmbH & Co. KG
Alle Rechte vorbehalten. / All rights reserved. Saarbrücken 2014

# Abstract

A polymer has drastically different physical properties above versus below some characteristic temperature. For this reason, the precise identification of this glass transition temperature, $T_g$, is critical in evaluating product feasibility for a given application.

The objective of this book is to review the behavior of polymers near their $T_g$ and assess the capability of predicting $T_g$ using theoretical and empirical models. It was determined that all polymers undergo structural relaxation at various temperatures both nearly above and below $T_g$, and that practical assessment of a single consistent $T_g$ is successfully performed through consideration of only immediate thermal history and thermodynamic properties. It was found that the best quantitative structure-property relationship (QSPR) models accurately predict $T_g$ of polymers of theoretically infinite chain length with an average error of less than 20 K or about 6%, while $T_g$ prediction for shorter polymers must be done by supplementing these $T_g(\infty)$ values with configurational entropy or molecular weight relational models. These latter models were found to be reliable only for polymers of molecular weight greater than about 2,000 g/mol and possessing a $T_g(\infty)$ of less than about 400 K.

# Table of Contents

# List of Figures

# List of Tables

# Acknowledgments

This book was made possible through the continuous support of the faculty and staff at *Kansas State University*. Special thanks goes to Dr. Jennifer Anthony for her support and constructive feedback throughout the composition of this book. I would also like to recognize Dr. Larry Glasgow, whose tremendous enthusiasm and vast breadth of knowledge in the engineering sciences served as a great source of inspiration.

Lastly, I would like to thank my family. The early encouragement and formative influence from my mother Mary Mlynarczyk to pursue a career in engineering has led me to where I am now, and Asima Ali continues to bring the best out in me every day.

# Introduction

The glass transition temperature, $T_g$, denotes the changeover point at which a material behaves like a glass or a rubber, and is among the most important characteristic values of a polymer. The drastic differences in physical and mechanical properties below versus above the $T_g$ make its precise identification critical, and is thus one of the first values measured after synthesizing a new polymer. The $T_g$ value will dictate the acceptable operating temperature range of a polymer for a desired application. In the context of synthesis of new polymer compounds, the $T_g$ will determine the associated feasibility for the intended application. This book aims to evaluate the capability of current theoretical and empirical models to characterize and predict the $T_g$.

An illustration of the potential catastrophic consequences of oversight of appropriate operating temperature range can be seen in the infamous space shuttle *Challenger* disaster. Rubber O-rings composed of fluoroelastomers were used as seals between two sections of the solid-fuel rocket boosters. The elastic property required for proper function of the O-ring was only present at temperatures above the $T_g$. Engineers at the time rated this safe operating threshold to be 40 °F, while the temperature prior to launch was only about 28 °F (Rogers Commission, 1986). Shortly after launch, the O-rings failed to flex and perform the proper seals, causing pressurized hot gas from the solid rocket motor to reach and impinge on the external fuel tank, thus leading to explosion of the vessel. This tragic disaster prompted significant reform in the testing of polymeric materials, and continues to serve as an engineering case study.

The experimental measurement of $T_g$ is performed accurately within a few degrees using one of several common laboratory methods. Differential scanning calorimetry (DSC), dynamic mechanical analysis (DMA), and thermomechanical analysis (TMA) are a few such techniques employed in industry. These techniques, although unique in experimental design and mechanism of action, all operate according to a similar template. Temperature is varied across a polymer sample, and an instantaneous spike in the value of a specific thermodynamic or physical property identifies the onset of the state transition, and thus determines the $T_g$.

The theoretical and computational prediction of the $T_g$, unlike experimental measurement, encompasses a multitude of approaches. Some models focus on the time-dependent structural relaxation mechanisms near the temperature of interest, while others instead rely on variables specific to the chemical structure of the polymer compound. Each model has its advantages and drawbacks, and often carries only selective applicability to certain classes of polymers. The proposal and refinement of such models, directed towards the goal of universal, fast, and reliable $T_g$ prediction remain an area of strong research interest (Le et al., 2012).

The forthcoming sections of this book discuss some of the prevalent models used to characterize and predict $T_g$. First, amorphous materials and the glass transition are defined in a broad sense. The kinetics of the glass transition are then described using existing correlation functions that seek to define the response behavior and structural relaxation mechanisms of polymers. After an overview of the common laboratory techniques used to measure $T_g$, some existing empirical models used to predict $T_g$ are introduced. The results obtained from these models are then assessed for accuracy and reliability by comparison with experimental values.

# Chapter 1 – The Glass Transition

The glassy state has many practical implications in a multitude of industries, ranging from food processing to biochemical stabilization. The understanding of the glass transition is essential in achieving mastery of efficient production and processing of polymeric materials and other amorphous solids.

Structural glass is a subset of glassy materials that refer to conventional amorphous solid materials with configuration disorder. By definition, glass is non-crystalline, yet possesses some of the same mechanical properties as crystalline solids (Lubchenko & Wolynes, 2007). Glass is also viewed as a vitrified form of a supercooled, extremely viscous liquid that does not undergo viscous flow or structural rearrangement on any observable timescale (Angell & Goldstein, 1986). Preparation of these materials occurs by rapid cooling of a molten liquid, of which the cooling rate has a significant impact on the final product properties. The cooling rate must be sufficiently high to prevent crystallization, yet variably low to achieve the desired mechanical properties (Moynihan et al., 1974). It has been found that for different cooling rates, microstructural changes are nearly negligible while mechanical and relaxational properties may vary significantly  (Painter & Coleman, 1997).

In measuring the value of a thermodynamic property $P$ such as enthalpy against temperature $T$ for a supercooled liquid undergoing a glass transition, various cooling rates yield different pathways. Thus, the temperature at which the given liquid experiences the transition is not static with respect to chemical structure and composition, but is instead dynamic and dependent on temporal effects.

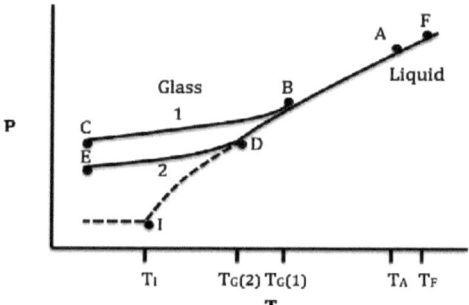

**Figure 1: Variation of a thermodynamic property P vs. temperature T for a typical supercooled liquid in the glass transition region (adapted from Fredrickson, 1988)**

In the above Figure 1, the liquid begins at state F and is supercooled to lower points on the curve. For a given cooling rate 1, the sample begins to transition to the glassy state at point B, completing the transition at point C along the diverted path as shown. However, choosing a slower cooling rate 2, the sample is able to remain in thermodynamic metastable equilibrium along the liquidus curve for longer, resulting in a lower glass transition temperature at point D. Both points B and D are nonequilibrium events termed laboratory glass transitions (LGT) (Gupta & Muaro, 2007). The theoretical minimum point at which the transition can occur by means of applying a minimum effective cooling rate is known as the ideal glass transition (IGT). The mechanism and ultimate existence of the IGT is often disputed in theory, with some researchers linking the IGT to either a thermodynamic or dynamical phase transition (Fredrickson, 1988). An underlying theme governing the latter transition type is that of ergodicity breaking, in which the time average behavior no longer coincides with the space averaged behavior (Palmer, 1982). This and other phenomena occurring near the glass transition temperature are subjects of active research in understanding the physical principles that govern the anomalous kinetics of structural glasses.

# Chapter 2 – Response Behavior

## 2.1 Linear Response Properties

The linear regime of the relaxation behavior of supercooled liquids occurs as result of relatively small perturbations from metastable equilibrium. Depending on how far away one is from the glass transition temperature LGT, the behavior mechanism of the relaxation can be quite different. At temperatures well above LGT, most materials exhibit simple single-exponential behavior illustrated by the Debye relation (Goldstein & Simha, 1976):

$$\varphi(t) = e^{-t/\tau} \tag{1}$$

where $\varphi(t)$ is the linear response function and $\tau$ represents a characteristic structural relaxation time, which at this condition is governed by an Arrhenius relation (Roland, 2008):

$$\tau = \tau_0 e^{E/k_B T} \tag{2}$$

where $E$ represents activation energy and $k_B$ is the Boltzmann constant. However, as the temperature is lowered near the LGT, the observed time-dependent structural relaxation becomes a nonexponential relation. An effective model for describing this modified behavior is known as the stretched exponential and is given by the Kohlrausch-Williams-Watts (KWW) function (Williams & Watts, 1970):

$$\varphi(t) = e^{-(t/\tau)^{\beta}} \tag{3}$$

where $\beta$ is simply $1/k_B T$ and is assumed to be less than 1.

A more fundamental view of Equation (3) can be expressed as (Debenedetti & Stillinger, 2001):

$$\varphi(t) = \frac{\sigma(t) - \sigma(\infty)}{\sigma(0) - \sigma(\infty)} \tag{4}$$

where $\sigma$ is the measured physical quantity. This stretched exponential function marks the existence of distinct spatially heterogeneous relaxing domains developing from the slowing down of long-time relaxation (Ediger, 2000). However, the analysis of these domains is limited since it is unclear whether they relax exponentially or nonexponentially.

Subdivisions of supercooled liquids are created from the characteristic temperature dependence and magnitude of parameters $\tau$ and $\beta$. Liquids with a high temperature independent activation energy and low temperature dependent $\beta$ value typically follow Equation (2) and are called strong liquids (Angell et al., 1986). Conversely, liquids with low activation energy and exhibiting non-Arrhenius behavior at low temperatures are termed fragile liquids (Angell et al., 1986). Some liquids exhibit properties of both strong and fragile liquids, and are deemed intermediate liquids (Böhmer et al., 1993). The temperature dependence of structural relaxation for these and fragile liquids can be expressed by the Vogel-Tamman-Fulcher (VTF) equation (Fulcher, 1925):

$$\tau = \tau_0 e^{E_0/(T-T_0)} \tag{5}$$

where $\tau_0$, $E_0$, and $T_0$ are material-specific parameters independent of temperature. By virtue of the asymptotic relationship of the effective activation energy $E(T)$ and temperature $T_0$, this relation assumes the existence of an IGT at $T_0$. As this assumption may not hold true for some liquids, the VTF equation should be selectively applied (Fredrickson, 1988).

An alternate expression for this temperature dependence of structural relaxation times is given by the Adam-Gibbs (AG) equation (Adam & Gibbs, 1965):

$$\tau = \tau_1 e^{E_1/T S_c} \tag{6}$$

where $S_c$ is configurational entropy and is a function of temperature.

In this model, the origin of viscous slow-down close to LGT lies in the decrease in the number of configurations in the system, and structural arrest is predicted to occur at a specific temperature. However, the concept of a cooperatively rearranging region (CRR) was used in the derivation of this expression (Adam & Gibbs, 1965). The temperature variation across the CRR determines the temperature dependence on relaxation behavior (Ngai et al., 1991). The weakness in this approach is the lack of definition of the size as well as the indistinguishable nature of this region, since stretched exponential behavior is believed to be governed by heterogeneity. Nevertheless, Equation (6) describes relaxational behavior for deeply supercooled liquids effectively (Angell & Smith, 1982).

Equation (6) is also related to a theoretical inconsistency known as the Kauzmann Paradox (Kauzmann, 1948). Specifically, extrapolations of $S_c$ to temperatures below LGT predict that $S_c$ disappears at some temperature $T_K$, which happens to be the predicted structural arrest temperature in the AG theory (Kauzmann, 1948). At temperatures below $T_K$, configurational entropy is considered negative, which would violate the third law of thermodynamics unless some phase transition were to occur. Therefore, the validity of Equation (6) also depends on the existence of an IGT. Equation (5) is obtained from Equation (6) if the difference in heat capacities between the supercooled liquid and its stable crystalline form are assumed to be inversely proportional to temperature (Goldstein & Simha, 1976). Thus, Equation (6) should be at least as applicable as Equation (5), with the likelihood that it can be applied effectively to a greater variety of materials.

## *2.2 Nonlinear Response Properties*

For large perturbations from metastable equilibrium, nonlinear time-dependent relaxation behavior is observed. In the nonlinear regime, the magnitude and sign of the perturbation affect the relaxation behavior. Temperature jump experiments reveal that relaxation from high and low temperature is asymmetric (Brawer, 1985). Specifically, for equivalent final temperatures and temperature jump values, an increase from a lower temperature will have a higher relaxation time than a decrease from a higher temperature (Fredrickson, 1988).

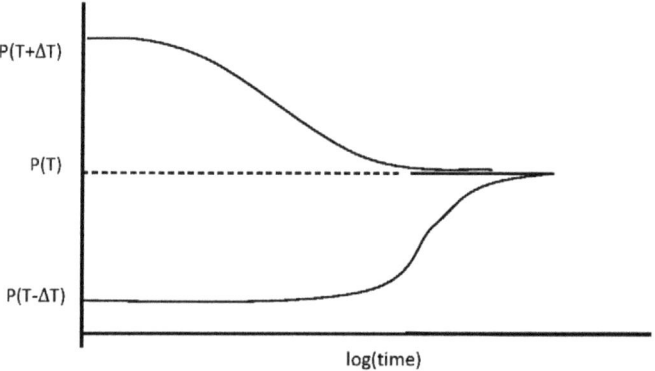

**Figure 2: Temperature jump experiments in a supercooled liquid revealing asymmetric nonlinear structural relaxation (adapted from Fredrickson, 1988)**

As seen in Figure 2 above, the final value of the temperature and thermodynamic property is the same in both experiments. In the experiment

6

that begins at the lower temperature T-ΔT, the time to reach the equilibrium P value after the +ΔT jump to T is considerably larger. The qualitative behavior seen in Figure 2 will be observed when ΔT is sufficiently large.

Further insight into the nonlinear regime of structural relaxation can be seen in experiments that involve multiple temperature jumps of varying sign, commonly known as "crossover" experiments (Fredrickson, 1988). In one variation of a "crossover" experiment, an initial temperature above the LGT is subject to a negative jump to a point below the LGT, but then quickly followed by a positive increase at about half the initial jump magnitude. The result is an unexpected overshoot in the equilibrium property value at the final temperature (Brawer, 1985). This overshoot is often explained by entropy changes. The sign of vibrational entropy contributions is a function of the quenching direction (Scherer, 1992). Specifically, this contribution decreases after down quenching due to the drop in temperature. In this case, the sign of the translational and vibrational contributions are opposite, which gives rise to an entropy maximum. The crossover effect is observed at some $T_x$ slightly greater than LGT, but only for appropriate magnitudes and kinetics of the various contributions (Brawer, 1985).

An alternate view reveals decoupling between translational diffusion and viscosity (Ediger, 2000) at a crossover point of $LGT < T_x < 1.2 \cdot LGT$. At this temperature, the inverse relationship between the translational diffusion coefficient and viscosity breaks down, although that for the rotational coefficient does not. A recent molecular dynamics (MD) simulation claims that the crossover effect can only be reproduced in simulations at sufficiently deep quenching temperatures and long aging times (Gupta & Muaro, 2007). Therefore, the study of crossover theory through MD is limited to these available conditions.

# Chapter 3 - Molecular Dynamics

Simulations known as molecular dynamics (MD) obtain useful physical information about a system by carrying out an integration of the equations of motion over hundreds of particles. Standard conventions adopted to investigate supercooled liquids are a single-component system, equivalency of compression and cooling, and defined scales for variables such as time and length (Barrat & Klein, 1991). The most commonly obtained dynamical quantity in MD, the self-diffusion coefficient $D$, can be calculated using either the Einstein (7a) or Kubo (7b) formulas (Wang & Hou, 2012):

$$D = \lim_{t \to \infty} \frac{1}{6t} \langle [r(t) - r(0)]^2 \rangle \qquad (7a)$$

$$D = \frac{1}{3} \int_0^\infty \langle v(t) \cdot v(0) \rangle \, dt \qquad (7b)$$

where $r$ and $v$ represent position and velocity vectors, respectively. In supercooled liquids, $D$ is typically small, primarily due to competing short-time and long-time mechanisms (Brawer, 1985). Thus, for supercooled liquids, the difference term in the mean squared displacement function offered in Equation (7a) is favorable over the velocity product of Equation (7b).

An extension of this expression leads to a parameter that can be interpreted as an order parameter for dynamic transition. Arising from the jump diffusion model, the Gaussian parameter can be written as (Barrat & Klein, 1991):

$$\alpha(t) = \frac{3}{5} \frac{\langle [r(t) - r(0)]^2 \rangle^2}{\langle [r(t) - r(0)]^4 \rangle} \qquad (8)$$

Deviations of $\alpha(t)$ from 1 indicate deviations from normal liquid behavior, and a stable non-unity value achieved at some time would indicate the presence of a new dynamic phase in the supercooled liquid (Bernu et. al, 1987). Obvious disadvantages to this MD method include the potential large timescale requirement, which may offset the convenience of the simple algorithm of the parameter.

Another analysis capability of interest in MD involves a spatial Fourier transform of the previously defined stretched exponential given by Equation (3). An expression known as the van Hove correlation function can be used to compute probabilities of finding particles at specific locations and times (Hopkins et. al, 2010). This expression can further be correlated with macroscopic hydrodynamics:

$$G_s(r,t) = N^{-1} \sum_{i=1}^{N} \langle \delta[\mathbf{r}_i(t) - \mathbf{r}_i(0) - \mathbf{r}] \rangle = \frac{1}{(4\pi Dt)^{3/2}} e^{-\left(\frac{r^2}{4Dt}\right)} \qquad (9)$$

The advantage of the latter expression in Equation (9) is that it gives rise to an opportunity to recognize structural arrest. Specifically, a plot of $4\pi r^2 G_s(r,t)$ vs. $r/\sigma$ drastically changes shape at some critical crossover density $n_x$ (Barrat et al., 1990). For $n > n_x$, diffusion seems to occur by neighbor jumps while for $n < n_x$ the diffusion behavior follows long-time hydrodynamics (Barrat & Klein, 1991). This approach proves useful in that the existence of a crossover point is clearly identifiable and can be fairly well fit to two distinct and comprehensible mechanisms. However, the precise quantitative point at which the transition occurs is difficult to identify with this method, as the precise time at which onset of a new curve shape occurs is subjective.

Other phenomena observed at the aforementioned crossover point prove useful in understanding the underlying physical mechanisms. While the transition is relatively broad, it can be more specifically marked by a change in the slope of the equation of state (EOS) (Bernu et al., 1987). Here, as the fluid undergoes structural arrest, a nonzero shear modulus also appears. The challenge in this approach is selecting the most appropriate EOS for the liquid. Once this challenge is met, this method is effective in providing more precise identification of the crossover point.

The appropriateness of approaches given by MD simulation may be system-specific, but the overall advantages and disadvantages of MD are summarized in Table 1:

**Table 1: Advantages and disadvantages of MD simulation as applied to supercooled liquids**

| Advantages | Disadvantages |
|---|---|
| - Short run-time<br>- Readily perform detailed calculations on simple particle interaction systems<br>- Given access to multiple correlation functions<br>- Simultaneously provide information on both structural and thermodynamic properties of a system | - Results quickly invalidated during ergodicity breaking phenomena<br>- Limited workable timescale<br>- Complications of component phase space for relaxation timescales greater than observed timescale<br>- Properties may become a function of thermal history and current external thermodynamic parameters |

# Chapter 4 – Structural Relaxation Models

## 4.1 Kinetic and Hydrodynamic Models

The extension of hydrodynamics and kinetics to supercooled liquids offers considerable insight into the structural arrest mechanism. By applying nonlinear theory and mode-coupling, Leutheusser proposed a time correlation function in the form of a nonlinear integro-differential equation (Leutheusser, 1984):

$$\ddot{C}(t) + \gamma \dot{C}(t) + \Omega^2 C(t) + \Omega^2 \int_0^t M(t - t') C(t') dt' = 0 \qquad (10)$$

Here, $C(t)$ represents a time correlation function and $M(t)$ represents a memory function, with $\gamma$ and $\Omega$ being damping and oscillation constants, respectively. By applying a low-order mode-coupling approximation to the memory function, Leutheusser was able to explicitly relate $C(t)$ and $M(t)$, thus making the differential equation solvable for $C(t)$ (Boone & Yip, 1991). The drawback to this approach, however, lies in the fact that it cannot be justified for timescales beyond the realm of high frequency expansions. Despite questionable validity, the Leutheusser model suggests that relaxation time follows a power law singularity and confirms the presence of the IGT (Fredrickson, 1988). This suggestion seems to follow experimental nonexponential behavior of some liquids (Taborek et al., 1986), and thus may possess some validity. The approach, however, is hindered because it neglects wave vectors related to density fluctuation. Since density fluctuations via a nonlinear feedback mechanism is the proposed driver of structural arrest in the model (Leutheusser, 1984), improved treatment of these vectors is essential for due diligence.

A better treatment of wave vectors using a similar approach to Leutheusser was performed by Bengtzelius and Kirkpatrick (1984), but arrived at many of the same results with only slightly modified exponents and relaxation spectrum broadness. These methods are also limited by the viscosity of the fluid, with fluids of viscosity beyond a certain threshold conflicting with molecular simulation results (Götze, & Sjögren, 1987) primarily due to the questionable mode-coupling theories that are used to relate $M(t)$ to $C(t)$.

A model proposed by Das et al. (1985) employs basic fluid mechanics equations in the framework of hydrodynamic theory. The hydrodynamic model possesses a pressure term, convective term, dissipative term, and Gaussian noise term with no structural order parameters. Upon selection of an appropriate potential energy function, the results of this approach share many similarities with Leutheusser, including the feedback mechanism. However, it also introduces other nonlinearities, which cause the IGT to vanish while still retaining many of its effects (Das et al., 1985).

A summary of hydrodynamic relative to kinetic models appears in Table 2:

**Table 2: A comparison of hydrodynamic and kinetic models for relaxation behavior of supercooled liquids**

| Advantages | Common Drawbacks | Disadvantages |
| --- | --- | --- |
| • Not restricted to fluid type<br>• Simple and precise<br>• Extendable to higher order fluids | • Unknown correlation to structural order parameters<br>• Questionable mode-coupling approximations | • Some parameters are found through questionable independent liquid theory equations<br>• Uncertainty of wave vector dependence |

## 4.2  Spin Models

Another approach in studying these liquid phase transitions is that of *n*-spin facilitated Ising models (nSFM) (Fredrickson & Andersen, 1984). In these models, spin-up is interpreted as a region of supercooled liquid with larger compressibility. With the imposition of a positive magnetic field, the number of these up-spins decreases with temperature. The flipping probability function is defined with high dependence on neighboring particles, thus leading to the theory of flipping by cooperative events, which becomes the proposed mechanism for relaxation (Fredrickson, 1988). One weakness of this approach is in the possibility of reducible dynamic constraints. In such a case, partitions would be necessary, which would lead to nonergodic behavior and thus alter the relaxation mechanism significantly. Therefore, the method essentially must depend on the appropriate restrictions.

The nSFM model can be reduced to specific choices of *n*. For example, the 1SFM model is used to represent isolated up-spins among a large number of down-spins. The immediate neighbor of the up-spin is allowed to flip-up while the original particle flips down, thus modeling a type of defect propagation analogous to a low-temperature relaxation mechanism (Fredrickson & Andersen, 1984). Monte Carlo (MC) simulations have confirmed that the 1SFM relaxation model obeys the Arrhenius expression of Equation (2) (Fredrickson, 1988). Overall, this model carries the advantage of being thermodynamically well-defined, but also relies heavily on a questionable spin-up conserving diffusion mechanism.

In the 2SFM view, surfaces of up-spins move in concert to relax surrounding down-spins through cooperative dynamics. Although perturbation theory predicted the existence of an IGT under this model, MC simulations have refuted this (Fredrickson, 1988). The simulations did, however, indicate nonexponential time decay and non-Arrhenius temperature dependence on relaxation (Angell & Goldstein, 1986). The results also indicate that 2SFM agrees well with the AG Equation (6), but cannot be extrapolated to lower temperatures due to entropy function curvature (Dorfmüller & Williams, 1987).

Through application of various lattices and spin models, KWW (3) and VTF (5) behavior can also be derived (Fredrickson, 1988). Thus, the nSFM model is excellent as a supplement to other relaxation models, but as a standalone model may be too variable with respect to dynamic parameters to distinctively explain the glass transition.

## 4.3   Square Tiling Model

In a model proposed by Weber, Fredrickson, and Stillinger (1986), a supercooled liquid is represented by an area of squares of varying sizes. In this Square Tiling Model (STM), each square represents a region of liquid containing well-packed molecules while each boundary represents regions of weakened bonds between these liquid sections. It is predicted that at the phase transition, these interior walls become unstable and expand, which reduces the system to a single square domain of dimensions $L \times L$ (Weber et al., 1986). This model is dependent on the selection of an appropriate potential energy function, and is also limited by two-dimensional dynamics. An advantage to this is that the system is well contained, with area conservation being a strict constraint. However, any possible three-dimensional dynamic behavior is lost in such a model since all changes in the state of the system must be represented with two-dimensional phenomena. A visual representation of a square tiling model can be seen in Figure 3 below.

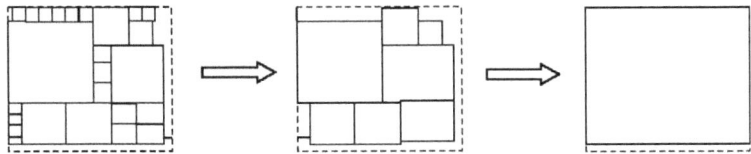

**Figure 3: A sample evolution in a square tiling model representing structural relaxation**

As seen in Figure 3, the boundaries of smaller squares collapse during relaxation to form larger squares. Relaxation is complete when only a single large square remains with a side length equal to the dimension of the defined system. Two sub-models exist in this view (Weber et al., 1986), and are compared in Table 3:

**Table 3: A comparison of square tiling sub-models using different kinetic rules**

| Model | Description | MC results | Disadvantages |
|-------|-------------|------------|---------------|
| Minimal aggregation | Square domains can fragment only if a dimensional condition is satisfied. Inverse aggregation is permitted. | - Relaxation occurs by KWW (3) behavior<br>- Arrhenius (2) and AG (6) are not satisfied<br>- Nonlinear phenomena | - Unconventional IGT with nonsingular relaxation times<br>- Arbitrary long-range constraints |
| Boundary shift | Square domain can fragment into domains of unit squares. Inverse shift is permitted. | - KWW (3) behavior<br>- Arrhenius (2) and AG (6) are not satisfied<br>- Faster relaxation than minimal aggregation | - Dependent on identical domain sizes<br>- Lattice spacing constraints |

# Chapter 5 – Experimental Measurement Techniques

Several laboratory techniques are available for the precise measurement of the glass transition temperature. The optimal technique is most often dependent on the physical properties and available sample volume of the compound to be measured.

## 5.1  Differential Scanning Calorimetry

A common thermoanalytical technique that can be applied to the identification and measurement of phase or state transitions such as the LGT is differential scanning calorimetry (DSC). In this technique, the temperature of experimental and reference samples is linearly increased and the corresponding amount of heat required is continuously measured. As the experimental sample undergoes a phase transition, more or less heat is required to maintain it at the same temperature as the reference, causing a spike to be observed on the recorded DSC signal. Specifically, in the case of the LGT, the sample undergoes a change in heat capacity even though no formal phase change occurs, and so the measured heat flow will experience a step increase at that temperature. Since this step generally occurs over the range of a few degrees, the LGT is taken to be the center point of the incline (Skoog, 1998).

Besides the state transition of the LGT, common phase transition temperatures measured via DSC include that of crystallization ($T_c$) and melting ($T_m$). Since crystallization is an exothermic process while melting is endothermic, the DSC signal experiences a negative and positive step, respectively, at these events. A typical DSC plot containing these transition points is shown in Figure 4. Values of $T_m$ in relation to $T_g$ for several common polymers are given in Appendix A.

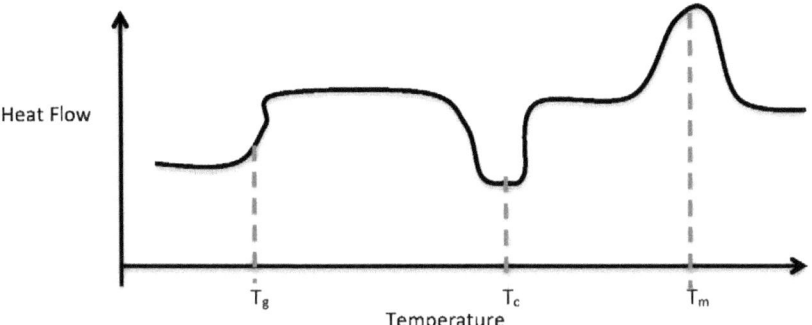

**Figure 4: A typical differential scanning calorimetry plot
with commonly identified state and phase transitions**

## 5.2 Dynamic Mechanical Analysis

A very common method used in the characterization of the viscoelastic
behavior of polymers is that of dynamic mechanical analysis (DMA). In this
technique, a sinusoidal stress is applied to the test specimen and the
magnitude and phase shift of the resulting strain is measured. Alternatively,
the converse procedure may be employed in which strain is the input and the
resulting stress is the measured output. With the gathered stress-strain data,
one can compute the storage (E') and loss (E'') moduli as follows (Meyers &
Chawla, 2010):

$$E' = \frac{\sigma_0}{\varepsilon_0}\cos{(\delta)} \tag{11}$$

$$E'' = \frac{\sigma_0}{\varepsilon_0}\sin{(\delta)} \tag{12}$$

where $\sigma_0$ is the stress magnitude, $\varepsilon_0$ is the strain magnitude, and $\delta$ is the
phase lag between stress and strain.

In order to discern the LGT, the sample temperature is varied and compared against the resulting moduli in what is known as temperature-sweeping DMA. At the LGT, a dramatic decrease in the storage modulus along with a maximum in the loss modulus is observed. As seen through the combination of equations (11) and (12), this also equates to a peak in the ratio of $E''$ to $E'$ or $\tan(\delta)$, known as the material loss factor or loss tangent.

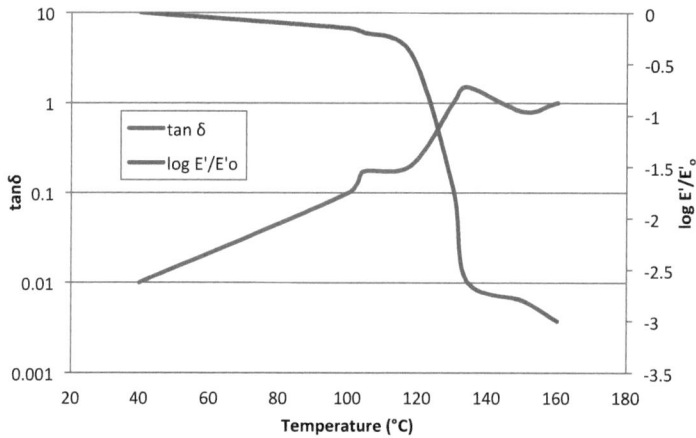

**Figure 5: Temperature-sweep dynamic mechanical analysis spectra of the material loss factor and storage modulus for a PC/ABS polymer blend (adapted from Más et al., 2001)**

While the LGT can be deduced from DMA data by either the peak $E''$ or peak $\tan(\delta)$ value, the latter is the more prevalent in literature. The peak $\tan(\delta)$ value is several degrees higher than the peak $E''$, and corresponds more closely to the transition midpoint as opposed to the onset of the state transition (Seyler, 1994). As is evident from Equations (11) and (12), the peak in $\tan(\delta)$ arises from a compromise between the $E''$ maximum and $E'$ minimum. In Figure 5 above, the LGT is identified as being located roughly at 134 °C.

## 5.3    Thermomechanical Analysis and Dilatometry

Just as in DSC, another technique that utilizes a linear temperature program is thermomechanical analysis (TMA). With this method, a constant stress is applied to the polymer sample and the resulting dimensional changes are measured. Although the magnitude is held constant, the applied stress may be implemented in one of several directions and configurations including compression, tension, flexure, and torsion. So naturally, TMA lends itself to multiple instrumentation configuration geometries and a high degree of flexibility in experimental design. The heat transfer in a TMA is considerably slower than in a DSC, so the heating rates are typically limited to about 10 °C/min (Seyler, 1994).

The special case in which a flat-tipped probe is used to measure the expansion in a single dimension is referred to as linear thermodilatometry, and is a common method employed by many laboratories in determining the LGT (Earnest, 1994). Dilatometry is a technique qualitatively very similar to TMA in that the dimensional changes of a material are measured against temperature. Dilatometers, however, are generally used to measure expansion in larger samples. In the dimension of interest, samples measured in a dilatometer are typically 25 times longer than those measured by TMA. While dilatometers are generally more stable and easily calibrated, TMAs are especially suitable for thinner polymer samples on the order of less than 0.1 mm (Seyler, 1994).

The primary variable responsible for the dimensional change incurred by a polymer at the LGT is the coefficient of thermal expansion (CTE). The value of the CTE in the glassy state is low, but the increased degree of segmental molecular motion in the rubbery state causes the CTE to be relatively high. Therefore, the slope of the dimensional change versus temperature curve experiences a sizable increase at the LGT.

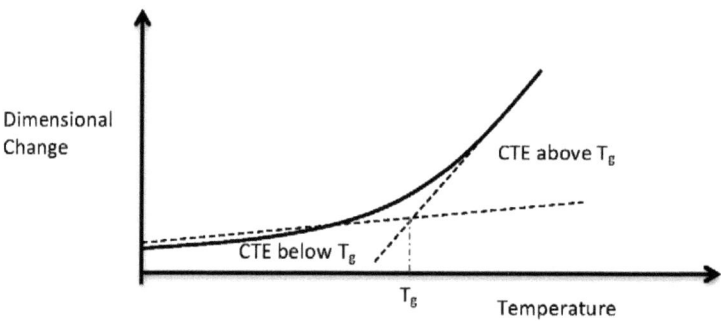

**Figure 6: A typical thermomechanical analysis curve featuring the extrapolation of the glass transition temperature from the glass and rubbery CTE domains**

While the transition as measured by the slope of the TMA curve may often be smooth, the LGT may still be measured accurately through the extrapolation of tangent lines. The intersection of these tangent lines representing the distinct linear domains serves as the approximation to the LGT, as illustrated in Figure 6 above.

## 5.4   *Thermo-optical Analysis*

A technique that relies more on visual observation is thermo-optical analysis (TOA). The polymer sample is subjected to a temperature program and the resulting light intensity is measured with a photocell. The physical property of birefringence, in which a material's refractive index is a function of light polarization and direction of propagation, encounters a drastic decrease at the LGT and thus serves as the basis of measurement. A significant advantage of this technique is in its ability to measure the LGT of very small samples, on the order of fractions of milligrams (Seyler, 1994). The primary disadvantage is in its inability to analyze transparent samples, which can be overcome by combining with DSC.

**Table 4**: Comparison summary of common laboratory $T_g$ measurement techniques

| Measurement Technique | Variable Exploited | Considerations |
|---|---|---|
| DSC | Heat Capacity | <ul><li>Relatively large range of heating rates available</li><li>Larger samples should be run at lower heating rates</li><li>Smooth baseline with minimal noise required for accuracy</li></ul> |
| DMA | Energy complex moduli | <ul><li>Good instrument temperature and force calibration required</li><li>Sample needs to have proper aspect ratio and even thickness</li><li>More sensitive than TMA</li></ul> |
| TMA | Coefficient of Thermal Expansion | <ul><li>Advantage: Best for measuring thin (< 0.1 mm) samples</li><li>Dilatometer used for longer samples</li></ul> |
| TOA | Birefringence | <ul><li>Advantage: Can measure very small samples</li><li>Disadvantage: Cannot analyze transparent samples</li><li>Can be combined with other techniques such as DSC</li></ul> |

In addition to these general considerations, the optimal technique for a given sample can be determined empirically. By recognizing specific cases where peaks are indiscernible, an alternate setup or different technique should be elected. It is also worthy to note that since $T_g$ occurs over a temperature range and these techniques are monitoring different processes, differences in measured $T_g$ value of a few degrees are common and expected.

# Chapter 6 – Polymer Glass Transition Temperature Predictor Models

As in supercooled liquids, the glass transition in polymers is similar to a second order phase transition. Thus, thermodynamic variables related to the second derivatives of the free energy will experience a discontinuity at the LGT (Van Dijk & Wakker, 1997). These include heat capacity, isothermal compressibility, and the CTE. Experimental techniques such as DMA, DSC, and TMA display this discontinuity in the form of a peak at the LGT. Such techniques are among the predominant methods for practical measurement of the LGT.

For design of new polymeric compounds and mixtures, however, it is useful to first predict the LGT prior to experimental measurement. Among other benefits, this allows for various potential product candidates to be surveyed prior to synthesis. As the desired application of the polymer often may involve temperatures very near to that of structural arrest and the LGT, fine resolution in accuracy is often required. Consequently, the development of simple and reliable LGT predictor models with wide applicability is of great value in industry.

Since polymer behavior is very much a function of thermal history, degradation effects from repeated application are a legitimate concern. When exposed to elevated temperatures near those that would constitute mechanical failure, polymers may undergo thermal degradation. In this process, the polymer is essentially fragmented into smaller molecules or monomers by one of several mechanisms including random scission, depolymerization, or side group elimination. However, although moderate changes in molecular weight commonly result, the LGT remains relatively insensitive to such changes (Crompton, 2010). As such, the consideration of only immediate thermal history along with base structural and thermodynamic features is generally sufficient to characterize properties near the LGT.

Studies of many polymer systems have shown that although properties of the glassy and transition regions are independent of chain length, the rubbery and liquid flow properties and their corresponding temperature range are markedly dependent on chain length (Bailey et. al, 1981). Besides chain length, various other structural properties of a given polymer play a key role in determining the temperature at which the glass transition will occur. While molecular weight is the central descriptor in many classic predictor models, variables that describe the chain stiffness and intermolecular forces in the polymer structure have also been found to profoundly impact the LGT. The appearance of specific constituent groups can also have a common influence, and often such variables prove to be more critical in certain classes of polymers over others.

## 6.1 Molecular Weight Relational Models

The earliest $T_g$ models investigated the relationships that exist among a few of the most basic thermodynamic variables in a polymer. The polymer is first represented as a function of its constituent parts as $A(X)_p B$, in which $A$ and $B$ are the end groups, $X$ is the repeating monomer unit, and $p$ is the number of these monomer units or degree of polymerization. The molar volume $V$ and molecular weight $M$ can then be expressed as a sum of the contributions from the $p$ monomer units and the end groups. Combining these two expressions through division leads to an equation for the specific volume for a polymer of degree of polymerization $p$:

$$v(p) = v(\infty) + \frac{V_e - m_e v(\infty)}{M} \tag{13}$$

where $v(\infty)$ is the limiting specific volume for a chain of infinite length, while $V_e$ and $m_e$ are the combined end chain molar volume and molecular weight, respectively.

Assuming a linear correlation between specific volume and temperature represented by coefficient $\alpha_p$, Equation (13) can be rewritten as (Fox & Loshaek, 1955):

$$v(p) = v_0(\infty) + \alpha_\infty T + \frac{(m + m_e)(\Delta \alpha T + \Delta v_0)}{M} \qquad (14)$$

where the subscript 0 represents a value extrapolated to a temperature of 0 K, $m$ is the molecular weight of a single monomer unit, and a $\Delta$ represents the difference of a value at $p = 1$ and $p = \infty$. This expression completes a characterization of the interdependent behavior of volume, temperature, and molecular weight for a homologous series of polymeric liquids. The required parameters are obtainable from the structure and the $v - T$ curves for both the liquid monomer and infinite length polymer.

To relate $v$ and $T$ at the glass transition state, it is helpful to recognize that the transition occurs when the polymer encounters a drop in internal mobility that is related to the cooling rate time scale. It has been shown that this mobility value, sometimes referred to as segmental jumping frequency, is critically dependent on the specific volume and the following linear relationship holds (Bueche, 1953):

$$v_g = v_g(\infty) - B\left[T_g(\infty) - T_g\right] \qquad (15)$$

where the subscript $g$ represents the glass transition, B is a constant, and $\infty$ again indicates the limiting value for the infinite chain polymer.

Evaluating Equation (14) at the glass transition in both the generic and $\infty$ limit and substituting into Equation (15) leads to an equation that relates the glass transition temperature $T_g$ to the molecular weight $M$ for a given polymer:

$$T_g = T_g(\infty) \left[ \frac{\alpha_\infty - B - \dfrac{(m + m_e)\Delta v_0}{T_g(\infty)M}}{\alpha_\infty - B + \dfrac{(m + m_e)\Delta \alpha}{M}} \right] \tag{16}$$

which can be rearranged and written as the Flory-Fox equation (Fox & Flory, 1950):

$$T_g = T_g(\infty) - \frac{K_g}{M} \tag{17}$$

where the empirical constant $K_g = (m + m_e)[\Delta \alpha T_g(\infty) + \Delta v_0]/(\alpha_\infty - B)$.

The constant $K_g$ is related to the free volume in the sample, which is essentially a measure of the mobility of a polymer chain relative to its surrounding chains. As a polymer is cooled towards the LGT, free volume decreases until it eventually reaches a critical minimum value in which its chains are not free to move into alternate conformations. By virtue of its placement in a polymer chain, the end groups will account for a significantly higher fraction of the free volume available in a polymer chain compared to individual monomer units. For this reason, Equation (17) is accurately applied to high molecular weight compounds while only selectively applied to lower molecular weight polymers where the end groups would have a more profound influence.

There also exists a critical molecular weight for entanglement, above which $T_g$ remains constant (Mark, 2004). Therefore, Equation (17) becomes theoretically invalid for molecular weights above this value in addition to being invalid below some lower bound. However, the molecular weight for entanglement is usually significantly large such that $T_g$ calculated from Equation (17) is approximately $T_g(\infty)$, thus eliminating any upper bound of applicability. As a result, the Flory-Fox equation maintains accurate applicability for a wide molecular weight range. Just in the pilot study conducted by Fox and Flory (1950), empirical fitting of measured specific volumes for $T_g$ determination in polystyrene was accurately performed in the molecular weight range of 2,970 to 85,000 (Fox & Flory, 1950). The parameters of Equation (17) for polystyrene as determined in the pilot study were $T_g(\infty) = 373$ K and $K_g = 1.2 \times 10^{-5}$ (Fox & Loshaek, 1955). DSC measurements performed on polystyrene samples of varying molecular weight, however, showed the value of $T_g(\infty)$ to be approximately 381 K (Claudy et al., 1983). Applying this corrected parameter, Equation (17) for polystyrene is written as:

$$T_g = 381 - \frac{1.2 \times 10^{-5}}{M} \tag{18}$$

The calculated $T_g$ values can then be compared to DSC measured values (Claudy et al., 1983) as shown in Table 5:

**Table 5: Measured and calculated $T_g$ values for polystyrene of various molecular weights using DSC and the Flory-Fox equation (Claudy et al., 1983)**

| $M_w$ $(g/mol)$ | $T_g$ $(K)$ measured | $T_g$ $(K)$ calculated |
|---|---|---|
| 650 | 265.5 | 160 |
| 800 | 279 | 201 |
| 2100 | 328 | 321 |
| 2850 | 343 | 335 |
| 4000 | 353.1 | 349 |
| 17,500 | 369 | 374 |
| 37,000 | 378 | 377 |
| 275,000 | 379 | 380 |
| 600,000 | 380.5 | 381 |

The values in Table 5 suggest that Equation (17) carries reasonable accuracy of less than 2.4% error for polystyrene of molecular weight above approximately 2,100 g/mol. At low $M_w$ values the end groups have a stronger influence and cause an overestimation of the sample free volume. In these cases, $K_g$ is overestimated and the resulting predicted $T_g$ value is lower.

A second polymer studied by Flory and Fox during the early formulation of Equation (17) was polyisobutylene. The parameter values as determined by Fox and Loshaek (1955) were $T_g(\infty) = 210$ K and $K_g = 0.3 \times 10^{-5}$. Comparing the calculated $T_g$ values using Equation (17) and these parameters against recent reported measured values using DMA (Kunal et al., 2008) identifies a similar minimum bound of validity as shown in Table 6.

**Table 6: Measured and calculated $T_g$ values for polyisobutylene of various molecular weights using DMA and the Flory-Fox equation (Kunal et al., 2008)**

| $M$ $(g/mol)$ | $T_g$ $(K)$ measured | $T_g$ $(K)$ calculated |
|---|---|---|
| 300 | 184.4 | 110 |
| 1100 | 191.1 | 183 |
| 2500 | 193.5 | 198 |
| 12,200 | 204.6 | 208 |

For polyisobutylene, an error of 4.2% is achieved at a number-averaged molecular weight of 1100 g/mol while a 2.3% error is present at 2500 g/mol. Thus, Equation (17) has been shown to be reasonably reliable for common polymers above a molecular weight threshold of approximately 2,000 g/mol.

From the original Flory-Fox equation came the rise of alternate molecular weight relational models, some of which are illustrated by the following equations (Fox & Loshaek, 1955, Dobkowski, 1982, and Ogawa, 1992):

$$\frac{1}{T_g} = \frac{1}{T_g(\infty)} + \frac{K_g}{T_g^2(\infty)M} \qquad (19)$$

$$T_g = T_g(\infty) - \frac{K_b}{M + B} \qquad (20)$$

$$\frac{1}{T_g} = \frac{1}{T_g(\infty)} + \frac{K_c}{M} \qquad (21)$$

$$\ln T_g = \ln T_g(\infty) - \frac{K_d}{M} \qquad (22)$$

$$T_g = T_g(\infty) - \frac{K_g}{\sqrt{M \cdot M_w}} \qquad (23)$$

where $K_b$, $K_c$, and $K_d$ are empirical constants and $M_w$ is the weight-average molecular weight, used to supplement the standard number-average molecular weight $M$.

The overall accuracies of each of these equations are considered to be comparable (Kim et. al, 2008), and each shares many of the same essential characteristics. Specifically, each contains one or more empirical constants that must first be determined for the given polymer. In this way, the equations become usable only after a precedent is set for a given homologous series of a specific polymer. To do this, appropriate experimental data, generally in the form of dilatometric or viscometric measurements, must first be gathered and fitted to determine empirical parameters. Only then can the $T_g$ of a varied molecular weight of an established polymeric compound be predicted. Therefore, for a polymer belonging to a newly discovered homologous series, the use of any of these equations offers no direct advantage relative to direct laboratory LGT measurement methods. Indirect advantages manifest when varied molecular weights within the given homologous series are being surveyed.

It is worthy to note that molecular weight relational models can also be applied to binary polymeric mixtures. Binary polymer mixtures are prevalent in industry, and are in fact often synthesized for the specific purpose of lowing the $T_g$. The secondary component in the mixture is commonly a diluent known as a plasticizer. This additive works to increase the free volume of the system and consequently lowers the $T_g$, thereby extending the rubbery regime to lower temperatures. Since such mixtures are thermodynamically miscible, they are also miscible on a molecular scale. The blend exhibits a single LGT at a temperature intermediate to that of the respective constituent polymers. As the weight fraction of each polymer is altered, a systematic shift in the LGT also follows. The behavior of the LGT can be approximated as a composition averaged inversed additivity with respect to the constituents, written as the Fox equation as follows (Van Dijk & Wakker, 1997):

$$\frac{1}{T_g} = \frac{w_1}{T_{g,1}} + \frac{w_2}{T_{g,2}} \tag{24}$$

where $w_1$ and $w_2$ are the weight composition fraction of the respective constituent polymers. Equation (24) is most commonly applied in practice to polymer blends and statistical copolymers with great accuracy (Hiemenz & Lodge, 2007).

## 6.2   Disorientation Entropy Model

An alternate view of the glass transition uses the thermodynamic concept of configurational entropy as a central basis. Specifically, glass formation was suggested to arise from the loss of configurational entropy in the system, described as the vanishing number of configurational states accessible to the fluid at low temperatures (Gibbs & Di Marzio, 1958).   The configurational entropy of a polymer system is relatively defined as:

$$S_c = S^{liquid} - S^{glass} \qquad (25)$$

where $S^{liquid}$ and $S^{glass}$ indicate configurational entropies of liquid and glass states, respectively. In determining $S_c$ for a given system, a reference zero state system is often used. This represents a "pure" or ideal polymer which lacks diluent molecules and possesses an ideal glass transition temperature, $T_{g0}$. Assuming the energy contribution from the vibration about the lattice sites is negligible, $S^{glass} = 0$ in the reference zero system. Making the further assumption that the dependence of the number of polymer molecules $n$ is equal on both the ideal and laboratory glass transition temperatures, $S^{glass} = 0$ in the real system as well.

However, unlike the reference zero system, the real polymer system possesses a heat capacity that is dependent on $n$. The corresponding expressions for $S_c$ in the reference zero and real polymer systems can then be respectively written as (Chow, 1980):

$$S_c(0, T) = \int_{T_{g0}}^{T} \Delta C_p(T')d\ln T' \qquad (26)$$

$$S_c(n, T) = \int_{T_g}^{T} \Delta C_p(n, T')d\ln T' \qquad (27)$$

where $\Delta C_p$ is the difference in heat capacity between the supercooled liquid and glass.

Approximating transition increments of isobaric heat capacity as being independent of both temperature and composition, $\Delta C_p(n, T') = \Delta C_p(T') = \Delta C_p$. The real laboratory and ideal glass transition temperatures can then be related by the expression (Chow, 1980):

$$\ln\left(\frac{T_g}{T_{g0}}\right) = -\frac{1}{\Delta C_p}[S_c(n,T) - S_c(0,T)] \qquad (28)$$

In proposing expressions for the configurational entropies, a model representation of the polymer liquid must first be chosen. In a given polymer solution exist solvent molecules and polymer molecules of varying size and chain configuration. These solutions preclude the standard conditions necessary for an entropy of mixing expression based on mole fractions of small molecules. In an alternative solution theory proposed by Flory and Huggins (1941), the polymer solution is represented as a collection of lattice sites that can be occupied by any of the dissimilarly sized individual polymer segments or solvent molecules. Expressions for entropy can then be derived from statistical mechanics as functions of lattice volume fractions (Van Dijk & Wakker, 1997).

Employing the lattice model of Flory and Huggins, the $S_c$ of a polymer can be considered to consist only of the polymer's disorientation entropy, such that $S_c(n,T) = S_{dis}$ and $S_c(0,T) = 0$. In this way, Equation (28) can be simplified to (Kim et. al, 2008):

$$\ln\frac{T_g}{T_{g0}} = -\frac{S_{dis}}{\Delta C_p} \qquad (29)$$

where the disorientation entropy is expressed as a function of degree of polymerization (Lee et al., 2007):

$$S_{dis} = \frac{k_B \gamma_{dis}}{r}\left[\ln p + (p-1)\ln\left(\frac{z-1}{e}\right)\right] \qquad (30)$$

where $\gamma_{dis}$ is a proportional constant representing the degree of disorientation, and $z$ is the lattice coordination number.

The combination of Equations (29) and (30) yields an expression that relates $T_g$ to $T_{g0}$:

$$T_g = T_{g0}\exp\left[-\frac{\gamma_{dis}R}{\Delta C_p}\left(\frac{\ln p}{p} + \left(\frac{p-1}{p}\right)\ln\left(\frac{z-1}{e}\right)\right)\right]$$ (31)

where $R$ is the ideal gas constant.

Since the ideal glass transition temperature $T_{g0}$ is difficult to determine, $T_g$ can instead be related to $T_g(\infty)$. The expression for $T_g(\infty)$ can be determined by taking Equation (31) to the $r \to \infty$ limit, which can then be divided from Equation (31) to obtain an expression for $T_g$ as a function of $p$ and constants associated with the given homologous series of polymers (Kim et al., 2008):

$$T_g = T_g(\infty)\exp\left[\frac{\gamma_{dis}R}{\Delta C_p p}\left(-\ln p + \ln\left(\frac{z-1}{e}\right)\right)\right]$$ (32)

Varying $\gamma_{dis}$ while utilizing arbitrary sample parameter values of $T_g(\infty) = 400\ K$, $\Delta C_p = 20\ J/mol \cdot K$, and $z = 12$ illustrates the strong dependence of $T_g$ on $\gamma_{dis}$:

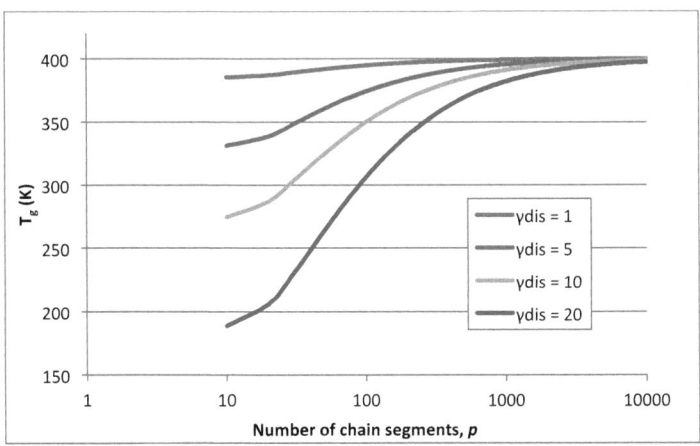

**Figure 7: Theoretical prediction of $T_g$ as a function of the number of chain segments for various degrees of disorientation using arbitrary constant sample parameter values (adapted from Kim et al., 2008)**

As the number of chain segments increases, the relative volume of chain ends decreases and so a reduction in free volume of polymer occurs. In this way, $T_g$ steadily increases with the number of chain segments. Eventually the value of $T_g$ plateaus near a critical molecular weight of entanglement, which corresponds to some high number of chain segments. Also seen in Figure 7 is the larger range in $T_g$ at higher values of $\gamma_{dis}$. Since a higher $\gamma_{dis}$ value represents a greater disorientation of polymer chains, a corresponding increase in disorientation entropy also follows. In effect, the polymer chains experience greater flexibility and mobility across all values of $p$, and a consequently lower $T_g$.

For the application of Equation (32) toward the prediction of $T_g$ for actual polymers, the value of $\gamma_{dis}$ was found for a set of five distinct polymers by nonlinear regression, and appears along with corresponding $\Delta C_p$ and $T_g(\infty)$ values in Table 7 below (Kim et al., 2008):

Table 7: Disorientation entropy model parameters for various polymers (Kim et al., 2008)

| Polymer | | $\Delta C_p$ $(J/mol \cdot K)$ | $T_g(\infty)$ (K) | $\gamma_{dis}$ | Monomer $M_w$ (g/mol) |
|---|---|---|---|---|---|
| Poly(α-methyl styrene) | PMS | 26.3 | 450 | 23.2 | 118.2 |
| Poly(methyl methacrylate) | PMMA | 32.7 | 385 | 9.7 | 100.1 |
| Poly(vinyl chloride) | PVC | 19.4 | 345 | 4.7 | 62.5 |
| Polypropylene | PP | 19.2 | 265 | 6.5 | 42.1 |
| Poly(dimethyl siloxane) | PDMS | 27.7 | 140 | 3.4 | 162.4 |

Using a lattice coordination number of $z = 12$ and comparing with DSC experimental data (Cowie, 1975), the continuous line curves plotted using Equation (32) are shown in Figure 8.

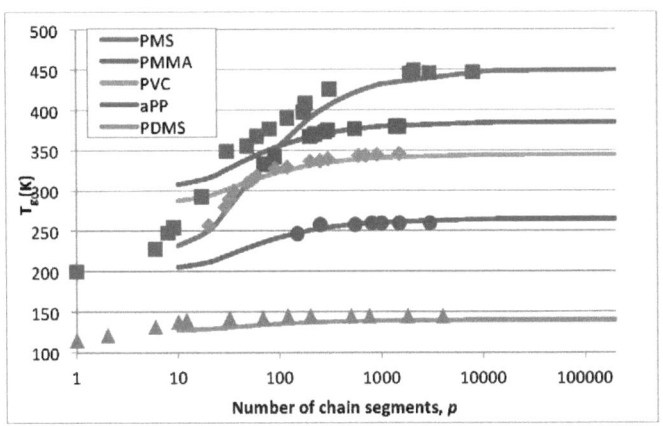

**Figure 8: Comparison of $T_g$ versus number of chain segments using disorientation entropy model (solid line) with reported DSC experimental values for various polymers (adapted from Kim et al., 2008)**

As can be seen in Figure 8, the range of applicability of the disorientation entropy model for the theoretical prediction of $T_g$ varies by polymer. For poly(dimethyl siloxane), the model seems to agree very well with experimental values all the way down to a molecular weight of about 1600 g/mol. On the other hand, the model $T_g$ values for poly(α-methyl styrene) agree with experimental values well only down to about 20,000 g/mol. As expected, the polymers with smaller $T_g$ range tend to have wider applicability with the disorientation entropy model. The data suggests that the model could be reliably used for polymer systems having $T_g(\infty) < 400\ K$ and greater than 100 chain segments.

## 6.3 Quantitative Structure-Property Relationship Models

Perhaps the most innovative approaches currently being taken in the prediction of glass transition temperature are those that focus on descriptors specific to the monomer. As the dependent variables are all related to the repeating unit structure, these models tend to be applied independently of polymer chain length (Katritzky et al., 1998). Thus, the polymers considered in such models are typically of large molecular weight past the critical value for entanglement. Essentially, these quantitative structure-property relationship (QSPR) models are estimating $T_g(\infty)$ for a given polymer, and can therefore be supplemented with molecular weight or configurational entropy models as an extension to other molecules in the given homologous series.

On the scale of a single polymer chain, the factors most affecting $T_g$ are chain stiffness and intermolecular forces (Mark, 2004). Stiffness of polymer chains is most significantly affected by the barrier of rotation around carbon-carbon bonds in the backbone chain, which is most influenced by the size of the substituent group bonded to these carbon atoms. When the backbone chain is allowed to rotate more freely, the $T_g$ is effectively lowered. For example, for a polymer of generic structural formula $-(CH_2 - CHR)_n -$ , a bulkier substituent group R yields a higher $T_g$ while a longer side chain serves to lower the $T_g$ (van Krevelen & Nijenhuis, 2009). Intermolecular forces are highlighted by the polarities of the repeating units and the hydrogen bonds that exist among the backbone chains and substituent groups. The $T_g$ is effectively increased by stronger attractive forces between backbone chains and larger polarity or charge-induced dipole of the side group that works to limit the free motion of the molecule.

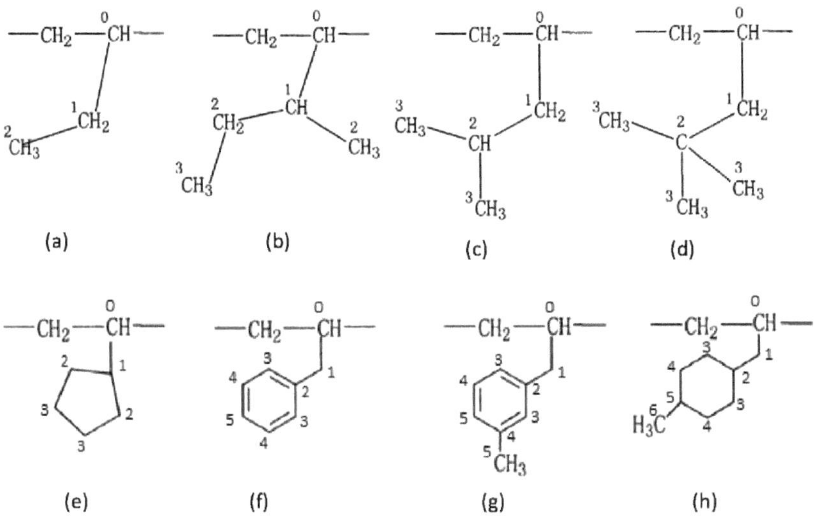

**Figure 9: Repeating unit structures of polyethylenes with side chain C atoms labeled by bond distance from backbone chain (adapted from Cao & Lin, 2003)**

It has been found that the size consideration that most affects the $T_g$ in polyethylenes is not the total size of the substituent group R, but is instead the size of the terminal group in the substituent group R (Cao & Lin, 2003). Therefore, parameters chosen for the chain stiffness variable of the model include the volume of the terminal group, $MV_{ter}$, in the substituent group R and the free length, $L_F$, of side chain (Cao & Lin, 2003).

As illustrated in Figure 9, the determination of lengths on the side chain is a straightforward task. The number above each carbon atom represents its minimum bond distance from the backbone chain carbon atom. To determine $L_F$, however, one must compute the length discrepancy between the side chain and its terminal group (van Krevelen & Nijenhuis, 2009). In polymer (a), the $^2C - {}^1C$ bond of the terminal group is able to rotate freely around the $^1C - {}^0C$ bond. Similarly, the terminal $^3C - {}^2C$ bond is free to rotate around the

$^2C - {}^1C$ bond in polymer (b). Polymers (c) and (d), on the other hand, contain multiple equivalent terminal groups. In polymer (c), it is necessarily true that if one of the two $^3C - {}^2C$ bonds rotates around the $^2C - {}^1C$ bond, the other $^3C - {}^2C$ bond does as well. Similarly, all three $^3C - {}^2C$ bonds in polymer (d) rotate simultaneously around the $^2C - {}^1C$ bond. If effect, the non-free rotation terminal parts of the side chains for polymers (a)-(d) are $-CH_3, -CH_3, -CH(CH_3)_2,$ and $- C(CH_3)_3$, respectively. The length discrepancy between the side chain and its corresponding terminal group can then be computed to determine the free length, $L_F$. Together with the cyclopentyl, phenyl, methyl-phenyl, and methyl-cyclohexyl terminal groups of polymers (e)-(h), the $L_F$ values for polymers (a)-(h) become 1, 2, 1, 1, 0, 1, 1, and 1, respectively (Cao & Lin, 2003).

Terminal group volume, $MV_{ter}$, can be calculated using one of a variety of molecular software packages. For a polymer $-(CH_2 - CR^1R^2)_n -$ containing two side groups $R^1$ and $R^2$ , it has been found that the resulting $T_g$ is lower in the case where the two side groups are equal, $R^1 = R^2$ , than if they are different (Cao & Lin, 2003). Therefore, the influence of the substituent group R on $T_g$ is dependent on molecular symmetry. The backbone chain of a symmetric substituted polymer seems to rotate more freely than for its asymmetrical counterpart. To capture the varied $T_g$ influence in the two cases, a sum volume of terminal groups can be used as the $MV_{ter}$ parameter for the asymmetrical substituted polymer while a margin volume can be used for the symmetrical case. With a regression fit, the two parameters $L_F$ and $MV_{ter}$ are sufficient for relatively accurate prediction of $T_g$ for nonpolar repeating units (Cao & Lin, 2003).

Monomer units that possess polarity require incorporation of parameters that capture the prevalent intermolecular forces. For a given polar repeating unit polymer, $-(CH_2 - CYZ)_n -$, the relative polarity of the monomer unit results from discrepancies in electronegativity between the Y and Z side groups as well as between the $CH_2$ and CYZ groups, along with polarizability effects of Y and Z (Bicerano, 2002). The presence of specific functional groups in Y and Z will also have marked effects on the main chains of the polymer. If an $-$ OH or $-NH$ group exists in the side groups, a hydrogen bond may be formed between polymer main chains. The existence of a $-C \equiv N$ group, on the other hand, would present an additional electrostatic attraction between main chains (Mark, 2004). Either of these added interactions would work to enhance the forces between the polymer backbone chains, thus limiting their ability to freely rotate. These intermolecular force effects can be effectively described by the introduction of three new parameters: the substituted backbone electronegativity discrepancy $\Delta X_{SB}$, the polarizability effect $\sum PEI$ of side group, and the electrostatic attraction $Q_{\pm}$ due to hydrogen bond between the polymer main chains (Cao & Lin, 2003).

Electronegativity discrepancy $\Delta X_{SB}$ for the $CH_2 - CYZ$ monomer unit can be calculated as the geometric mean of $|X_Y - X_Z|$ and $|X_{CH_2} - X_{CYZ}|$, expressed as (Katritzky et al., 1998):

$$\Delta X_{SB} = \sqrt{|X_Y - X_Z| \cdot |X_{CH_2} - X_{CYZ}|} \tag{33}$$

where any given group electronegativity $X_{eq}$ is computed via an equalization method using Pauling electronegativity units (Bratsch, 1984):

$$X_{eq} = \frac{\sum_i v_i + q}{\sum_i \frac{v_i}{X_i}} \tag{34}$$

where $v_i$ is the number of $i$ atoms, $q$ is the overall integral charge of the group, and $X_i$ is the initial pre-bonded electronegativity of atom $i$. The polarizability effect is computed as a sum over individual essential unit polarizability values and bond angles via the following expression (Chenzhong, C., and L. Zhiliang, 1998):

$$\sum PEI = \sum \left( \sum \frac{\alpha_i}{\left[ N_i \frac{1 + cos\theta}{1 - cos\theta} - \frac{2cos\theta(1 - cos^{N_i}\theta)}{(1 - cos\theta)^2} \right]^2} \right) \tag{35}$$

where $\alpha_i$ is the polarizability of the $i$th essential unit in the substituent, $N_i$ is the carbon atom number from the point charge $q$ to the $i$th essential unit, and $\theta$ is the $\angle CCC$ bond angle supplement. Atomic values for $\alpha_i$ , much like atomic electronegativity values $X_i$ , are readily available in literature (Haynes, 2011):

**Table 8: Atomic polarizabilities and Pauling electronegativities**

**of common atoms**

| Atom: | H | C | N | O | F | Cl | Br | I | S | P |
|---|---|---|---|---|---|---|---|---|---|---|
| $\alpha_i$ $(10^{-24}$ cm$^3)$ | 0.667 | 1.76 | 1.10 | 0.802 | 0.557 | 2.18 | 3.05 | 5.34 | 2.90 | 3.63 |
| $X_i$ | 2.20 | 2.55 | 3.04 | 3.44 | 3.98 | 3.16 | 2.96 | 2.66 | 2.58 | 2.19 |

The final intermolecular force parameter, the electrostatic attraction $Q_\pm$ due to hydrogen bond, is computed in cases where a hydrogen bond exists between the main chains, such as when a side group contains an –OH or -NH. The value of $Q_\pm$ is computed as a product of part charges on the two atoms again using Pauling units and the equalization method. For the generic bond $-MH$ in a side group, the expression becomes (Bratsch, 1984):

$$Q_\pm = q_M q_H \tag{36}$$

where partial charge $q_i$ is equal to $v_i(X_{eq} - X_i)/X_i$. As seen in Appendix B, a majority of polymers do not have a hydrogen bond connecting main chains (Katritzky et al., 1998), and thus have $Q_\pm = 0$. For polymers that do contain the hydrogen bond between main chains, $Q_\pm$ possesses a negative value.

40

The five molecular descriptors that serve as parameters can be summarized by the variables $\sum MV_{ter}(R_{ter}), L_F, \Delta X_{SB}, \sum PEI$, and $Q_{\pm}$. Through interrelation analysis, it was found that these five parameters are all significant descriptors in the model and are independent of each other, and the correlation with $T_g$ produces a first order regression equation (Cao & Lin, 2003):

$$T_g\ (K) = 203.97(\pm 5.58) + 0.39(\pm 0.03) \sum MV_{ter}(R_{ter})$$

$$- \ 8.93(\pm 0.90)L_F + 138.82(\pm 12.33)\Delta X_{SB}$$

$$- \ + 9.01(\pm 2.18) \sum PEI - 1174.41(\pm 216.89)Q_{\pm} \qquad (37)$$

In developing this model, a training set of 22 linear polymers of medium molecular weight were used (Cao & Lin, 2003). After successful formulation of the regression equation constants, the $T_g$ of other polymers can be predicted using Equation (37). Compared with experimental data for a set of 88 diverse polymers (Katritzky et al., 1998), the statistical $R^2$ value for the fit was 0.9056 with a standard deviation of 20.86 K and absolute average error of 15.30 K (Cao & Lin, 2003). This indicates a reasonably good correlation for the model with the prediction of $T_g$, but with significant issues in reliability for select polymers. For example, the predicted $T_g$ for poly(3,3-dimethylbutyl methacrylate) was 377 K, while the experimental value was only 318 K, producing an error of 18.6%. A complete listing of data for all 88 polymers can be found in Appendix B. Unfortunately, no common distinctive feature among the poorly estimated polymers could be identified, and the scatter plot shown in Figure 10 below reveals a relatively uniform error distribution.

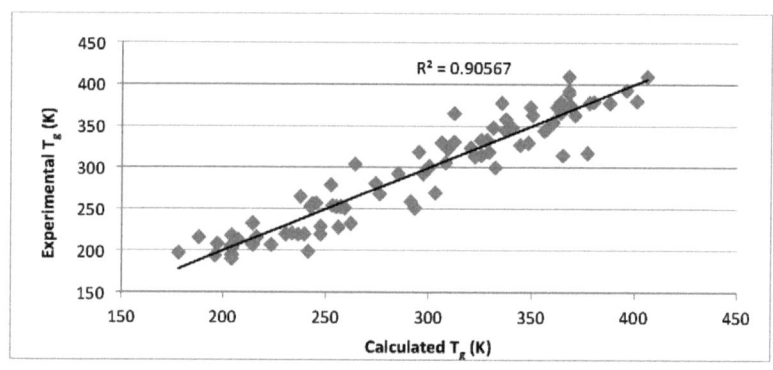

**Figure 10: Experimental versus calculated $T_g$ values using**

**5-descriptor QSPR model**

By employing the same template of a model driven by monomer unit descriptor parameters, many similar QSPR models have been derived and applied with comparable results. Models derived in current research typically rely more heavily on molecular software packages to extract and calculate the relevant molecular descriptor variables. Through the use of computational techniques such as density functional theory (DFT), monomer structures can be optimized for analysis during modeling (Katritzky et al., 1998).

In a recent study, a total of 1,664 molecular descriptors were calculated for each of 105 polyacrylate and polyvinyl molecules using DFT (Yu, Yu, & Wang, 2009). Multiple linear regression (MLR) with a training set of 50 experimental $T_g$ values was then used to seek an optimum subset of descriptors for incorporation into the model. For simplicity and robustness, only a few descriptors were sought for complete characterization of $T_g$. Ultimately, the optimal MLR model contained just three descriptor parameters, with physical meanings comparable to that of the 5-descriptor model previously discussed. Specifically, the descriptors used were mean atomic van der Waals volume, $Mv$, bond information content, $BIC5$, and electron diffraction 3D structure representation, $Mor13m$. Similar to the parameters of the previous model, these variables work to describe the chain stiffness and molecular mobility of the polymer. A statistical fit produced a corresponding regression equation (Yu, Yu, & Wang, 2009):

$$T_g = 73.050 + 698.016Mv - 278.545BIC5 - 54.569Mor13m \qquad (38)$$

The statistical $R^2$ value for the fit was only 0.861 with a standard deviation of 20.9 K, and the absolute error for the test set was approximately 21.7 K. Some improvement was achieved through the use of an alternate method to MLR known as an artificial neural network (ANN), in which the test set error was reduced to 17.7 K (Yu, Yu, & Wang, 2009). The overall results using this model were comparable to those seen in the previous model, with complete input parameter and output $T_g$ values tabulated in Appendix C.

Table 9: Comparison of predicted $T_g$ values obtained from 5-descriptor

and 3-descriptor QSPR models for various polymers

([a]Cao & Lin, 2003 and [b]Yu, Yu, & Wang, 2009)

| Polymer | $T_g$ (K) $calc^a$ | $T_g$ (K) $calc^b$ | $T_g$ (K) |
|---|---|---|---|
| | 5-descriptor | 3-descriptor | experimental |
| Poly(1-heptene) | 230 | 224 | 220 |
| Poly(3-methyl-1-butene) | 309 | 395 | 323 |
| Poly(3-pentyl acrylate) | 245 | 271 | 257 |
| Poly(3-phenyl-1-propene) | 325 | 314 | 333 |
| Poly(4-methyl-1-pentene) | 300 | 303 | 302 |
| Poly(acrylic acid) | 380 | 401 | 379 |
| Poly(ethyl acrylate) | 259 | 232 | 251 |
| Poly(methyl acrylate) | 274 | 312 | 281 |
| Poly(sec-butyl acrylate) | 255 | 237 | 253 |
| Poly(tert-butyl acrylate) | 325 | 325 | 315 |
| Poly(vinyl acetal) | 360 | 359 | 355 |
| Poly(vinyl acetate) | 300 | 292 | 301 |
| Poly(vinyl chloroacetate) | 264 | 305 | 304 |
| Poly(vinyl n-butyl ether) | 233 | 226 | 221 |
| Poly(vinyl n-octyl ether) | 196 | 218 | 194 |
| Poly(vinyl n-pentyl ether) | 223 | 218 | 207 |
| Poly(vinyl sec-butyl ether) | 242 | 245 | 253 |
| Poly(vinyl trifluoroacetate) | 295 | 294 | 319 |

As shown in Table 9 and Figure 11, significant differences in the predicted $T_g$ values exist between the two QSPR models. The $T_g$ value for poly(3-methyl-1-butene) is predicted with significantly greater accuracy using the 5-descriptor model (4.3% error) versus the 3-descriptor model (22.3% error), but the $T_g$ for poly(vinyl chloroacetate) is better predicted using the 5-descriptor model (0.3% versus 13.2% error). Since these models are constructed via regression and use different parameters, accuracy differences will exist for many polymers without easily identifiable molecular justification. The ratio of the test set to training set of polymers was 4:1 and 2.1:1 for the 5-descriptor and 3-descriptor models, respectively. Therefore, since more of the predictor set was used to develop the model regression equation, the 3-descriptor model equation has an advantage in relative accuracy. The disadvantage resulting from the fewer number of descriptors works to offset this advantage and make the two models rather competitive in accuracy.

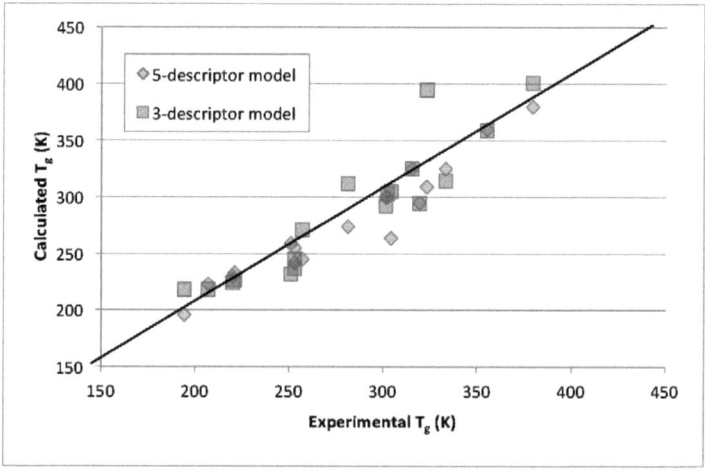

**Figure 11: Calculated versus experimental $T_g$ values for various common polymers using 3-descriptor and 5-descriptor QSPR models**

As seen in Figure 11 above, the largest differences between the two QSPR models for the common polymers surveyed exist in the 280-305 K experimental $T_g$ range. In this range, the 5-descriptor model tended to more greatly underestimate the $T_g$ than did the 3-descriptor model. While some specific conclusions on the relative performance of the two models can be drawn for this specific set of polymers, it would be expected that a different set of polymers would yield significantly different results.

# Chapter 7 – Discussion and Conclusions

The recent strides towards understanding the glass transition in supercooled liquids have been among the most significant in the history of the subject. The capabilities lent by molecular simulation have been proven to support proposed theories concerning the mechanism of the crossover point in these liquids, and continue to be invaluable in this area of research.

In the literature, the glass transition has been shown to be dependent on temporal effects. Studies have shown that for a polymeric liquid starting at some temperature in the rubbery regime, the cooling rate will impact the precise value of $T_g$. The minimum possible $T_g$, referred to as the ideal $T_g$ or IGT, has often been disputed in theory. Ultimately, research has been inconclusive as to whether a precise and unique IGT exists for any given polymer, and if it does exist the associated mechanism is also uncertain.

The current state-of-the-art models on glass transition include various linear and nonlinear relaxation correlations, which include those presented as Debye, KWW, Arrhenius, VTF, and AG in Equations (1) through (6). These response theories have been coupled with various dynamic models in the literature to describe structural relaxation. Many of these models have been shown to possess adequate simplicity to be solved effectively by analytical or simulation methods. MD simulations have been used in several studies to obtain useful results. These studies have been able to identify the point at which structural arrest seems to occur, but only within a reportedly broad range.

In other studies, kinetic and hydrodynamic models have been constructed through the application of theory involving principles such as mode-coupling, nonlinearity, wave vectors, and fluid mechanics. The Leutheusser and Bengtzelius & Kirkpatrick models both concluded that the unique IGT exists and relaxation time follows a power law singularity, but the Das model differed by concluding that an IGT does not exist.

Another work presented nSFM models as a means of describing the structural relaxation mechanism. In this family of models, polymer particles encountered flipping from a spin-up to a spin-down state with increasingly higher probability as temperature decreased. Supporting studies used MC simulations to conclude that nSFM models indicate nonexponential time decay and non-Arrhenius temperature dependence on relaxation. These MC simulations also concluded that the IGT does not exist in this model, although a separate study using perturbation theory indicated that the IGT does exist under the nSFM model.

In another study, MC simulations were also applied to an STM model. This two-dimensional model represented a polymeric liquid as an area of squares whose individual boundaries become unstable and vanish at decreasing temperatures. The MC simulations reportedly confirmed the existence of the IGT and showed structural relaxation to follow nonexponential behavior as illustrated by the KWW relation in Equation (3).

The results of many of these models don't agree on some key aspects. In studying all of the mentioned models, one is still not certain of the presence of an IGT, as models such as the STM reveal an IGT while those such as 2SFM or hydrodynamic models proposed by Das refute the presence of an IGT yet reveal some of the same characteristics associated with the phenomenon. The validity of some of the models is also questionable. For example, the simplifying assumptions made in the Leutheusser and Bengtzelius & Kirkpatrick models used mode-coupling approximations that are not justifiable at all timescales. The nSFM model relies on a questionable spin-up conserving diffusion mechanism and the STM model ignores three-dimensional phenomena. With any model of a desired simplicity, however, these types of disputable simplifying assumptions are bound to exist. In the end, these models are still able to provide considerable insight into the behavior of polymers near their $T_g$. The primary consensus that can be reached from all these theoretical models is the existence of a temperature-dependent structural relaxation mechanism, of which the physical particulars are likely fluid-specific and involving multiple phenomena.

Much further work is needed to advance the subject of the glass transition. Primarily, a better understanding of the temperature region between the crossover temperature and the glass transition temperature must be achieved through better characterization of atomic motion, testing of multiple mode-coupling theories quantitatively, and also the testing of spin and square-tiling models quantitatively. Also, from the hydrodynamic end, the testing of non-Newtonian equations of motion may be appropriate for short-time dynamics. Hydrodynamic models that also incorporate molecular structure could be introduced and analyzed by MD to identify structural order parameters.

The relation between multiple relaxation behavior models provides endless research opportunities. For example, a connection between the hydrodynamic and spin models could aid in a construction of a hybrid hydrodynamic model with long-wavelength, low-frequency dynamics. Proving that the AG relation (6) holds for all nSFM models can be achieved by simulations with $n > 2$. Also, expanding on kinetic rules in the Square-Tiling model may give rise to other fragmentation mechanisms that translate to alternate modes of relaxation. Perhaps even application of pressure can introduce shifts in the crossover point and thus be incorporated into some future models.

Apart from understanding the relaxation behavior of polymers, the ability to predict $T_g$ quickly and reliably is of great interest. The primary basis of comparison is with the results obtained from experimental measurement. Novel empirical models that employ QSPR theory have been shown to have considerable accuracy and a great potential for improvement.

The experimental techniques described in the literature as being most utilized in industry all share a common template. The value of a physical or thermodynamic variable known to drastically change at the $T_g$ is measured across temperatures that span the $T_g$, and the observed spike is recorded and identified with the $T_g$. While it has been shown that most polymers can be accurately analyzed by any of the techniques discussed, limiting factors such as sample size and transparency naturally favor some techniques over others. For example, TOA was described as being best capable of measuring $T_g$ for very small samples but unable to analyze transparent samples. TMA was shown to be better suited to thinner samples while dilatometry was described as being applicable for larger samples. Since $T_g$ occurs over a temperature range and different experimental techniques monitor different thermodynamic processes, measurement variability of a few degrees between different techniques is common.

Several $T_g$ predictor models have been well described through previous studies. The early models were molecular weight relational and were derived using basic relationships between volume, molecular weight, and temperature. Equations (17) – (23) summarize this class of models, which all report to have fairly good accuracy. Another class of predictor models uses configurational entropy as a central basis. One example of such a model used parameters related to disorientation entropy to predict $T_g$. The published results showed that accuracy was generally very good for polymers possessing a large number of chain segments.

The $T_g$ predictor models published most recently were generally of the QSPR type. The parameters in these models were illustrated to be a function of the monomer unit, and so the models were said to be only applicable to polymers of large molecular weight. Using modern computation techniques, a large number of potential molecular descriptors were surveyed and effectively chosen via multiple linear regression or more advanced algorithms. The underlying phenomena of the relevant variables used in such models were related to chain stiffness and intermolecular forces within a given monomer unit. The accuracy of these advanced models have been reported to be on the order of about 20 K.

While the physical properties and characteristic temperatures of a polymer may vary significantly with varied applied cooling rates, practical considerations are typically performed at static operating temperatures. As the polymer possesses consistent behavior in isothermal and isobaric operation, a single $T_g$ can be effectively evaluated. While multiple laboratory techniques are readily available to accomplish this measurement for a given polymer sample, models to predict $T_g$ for new polymers are useful in surveying potential products for a given application. Selection of the appropriate model is first and foremost dependent on experimental precedent. Should varying chain lengths within a given established homologous series of polymer be considered, molecular weight relational models illustrated by Equations (17)-(23) or configurational entropy models such as that represented by Equation (32) are most appropriate. The necessary input variables to these model equations such as $T_g(\infty)$ and other polymer-specific constants must first be established through an experimental precedent. The reliability of these models is generally good to a lower bound of about 2,000 g/mol of molecular weight or a couple hundred chain lengths, and a maximum $T_g(\infty)$ of about 400 K.

For new polymers without an experimental precedent, quantitative structure-property relationship (QSPR) models that focus on structural characteristics of the monomer are most appropriate. Properties related to the terminal groups in these repeating units have proven to be key in predicting $T_g$, with bulkier terminal groups and lower free chain length raising $T_g$. Larger intermolecular forces arising from polarity or strong hydrogen bonding between main chains have also been found to serve as a barrier to free rotation and thus also increase $T_g$. Overall, QSPR models have proven to correlate well with experimental $T_g$ data, the better models achieving a statistical $R^2 > 0.90$ and average absolute errors of less than 20 K, or approximately 6%.

For $T_g$ predictor models, given that the variability in experimental measurement is as high as 5-10 K, absolute errors of less than 20 K are relatively satisfactory. The focus must be on achieving this level of accuracy across all polymers, as some polymers have shown to be statistical outliers with regards to measured accuracy. The reliability of the model regression equations can likely be increased by using a larger training set of more diverse polymers. The number of potential forms of molecular weight relational model equations have nearly been exhausted, all showing good mean accuracy above a certain chain length number threshold. Greater opportunity for improvement exists in the refinement of configurational entropy models, as a wide array of factors that may affect entropy exist as potential variable candidates. Prediction of $T_g$ for polymers of low molecular weight remains the biggest challenge.

While the average performance of the most novel QSPR models is satisfactory, refinement is necessary to also address certain polymer cases that yield larger errors. Perhaps an advanced multiple linear regression analysis of molecular characteristics of polymers yielding high $T_g$ value errors could illuminate additional relevant variables. Sacrificing simplicity may be appropriate, as extending models to incorporate additional molecular descriptor variables may yield improved accuracy.

# References

Adam, Gerold, and Julian H. Gibbs. "On the Temperature Dependence of Cooperative Relaxation Properties in Glass-Forming Liquids." *The Journal of Chemical Physics* 43.1 (1965): 139.

Angell, C. A., and Martin Goldstein. *Dynamic Aspects of Structural Change in Liquids and Glasses*. Vol. 484. New York, NY: New York Academy of Sciences, 1986.

Angell, C.A. and D.L. Smith. "Test of the Entropy Basis of the VTF Equation: Dielectric Relaxation of Polyalcohols near Tg." *Journal of Physical Chemistry*, 86, 3845 (1982).

Bailey, R. T., Alastair M. North, and R. A. Pethrick. *Molecular Motion in High Polymers*. Oxford: Clarendon, 1981.

Barrat, J.L. and Klein, M.L. "Molecular Dynamics Simulations of Supercooled Liquids near the Glass Transition." *Annual Review of Physical Chemistry*, 42, 23 (1991).

Barrat, J.L., J.N. Roux, and J.P. Hansen. "Diffusion, Viscosity and Structural Slowing down in Soft Sphere Alloys near the Kinetic Glass Transition." *Chemical Physics* 149.1-2 (1990): 197-208.

Bengtzelius, U., W. Gotze, and A. Sjolander. "Dynamics of Supercooled Liquids and the Glass Transition." *Journal of Physics C: Solid State Physics* 17.33 (1984): 5915-934.

Bernu, B., J. Hansen, Y. Hiwatari, and G. Pastore. "Soft-sphere Model for the Glass Transition in Binary Alloys: Pair Structure and Self-diffusion." *Physical Review A* 36.10 (1987): 4891-903.

Bicerano, Jozef. *Prediction of Polymer Properties*. New York: Marcel Dekker, 2002.

Böhmer, R., K. L. Ngai, C. A. Angell, and D. J. Plazek. "Nonexponential Relaxations in Strong and Fragile Glass Formers." *The Journal of Chemical Physics* 99.5 (1993): 4201-209.

Boon, Jean-Pierre, and Sidney Yip. *Molecular Hydrodynamics*. New York: Dover Publications, 1991.

Bratsch, Steven G. "Electronegativity Equalization with Pauling Units." *Journal of Chemical Education* 61.7 (1984): 588-89.

Brawer, Steven. *Relaxation in Viscous Liquids and Glasses: Review of Phenomenology, Molecular Dynamics Simulations, and Theoretical Treatment*. Columbus, OH: American Ceramic Society, 1985.

Bueche, F. "Segmental Mobility of Polymers Near Their Glass Temperature." *The Journal of Chemical Physics* 21.10 (1953): 1850.

Cao, C., and Y. Lin. "Correlation between the Glass Transition Temperatures and Repeating Unit Structure for High Molecular Weight Polymers." *Journal of Chemical Information and Modeling* 43.2 (2003): 643-50.

Chenzhong, C., and L. Zhiliang. "Molecular Polarizability. 1. Relationship to Water Solubility of Alkanes and Alcohols." *Journal of Chemical Information and Modeling* 38.1 (1998): 1-7.

Chow, T. S. "Molecular Interpretation of the Glass Transition Temperature of Polymer-Diluent Systems." *Macromolecules* 13.2 (1980): 362-64.

Claudy, P., J. M. LeToffe, Y. Camberlain, and J. P. Pascault. "Glass Transition of Polystyrene versus Molecular Weight." *Polymer Bulletin* 9-9.4-5 (1983): 208-15.

Cowie, J.M.G. "Some General Features of Relations for Oligomers and Amorphous Polymers." *European Polymer Journal* 11.4 (1975): 297-300.

Crompton, T. R. *Thermal Stability of Polymers*. Shrewsbury, U.K.: Smithers Rapra, 2012.

Das, Shankar, Gene Mazenko, Sriram Ramaswamy, and John Toner. "Hydrodynamic Theory of the Glass Transition." *Physical Review Letters* 54.2 (1985): 118-21.

Debenedetti, P. and Stillinger, F. "Supercooled Liquids and the Glass Transition". *Nature*, 410, 259 (2001).

Dobkowski, Z. "Influence of Molecular Weight Distribution and Long Chain Branching on the Glass Transition Temperature of Polycarbonate." *European Polymer Journal* 18.7 (1982): 563-67.

Dorfmüller, Thomas, and G. Williams. *Molecular Dynamics and Relaxation Phenomena in Glasses: Proceedings of a Workshop Held at the Zentrum Für Interdisziplinäre Forschung Universität Bielefeld, Bielefeld, FRG, November 11-13, 1985*. Vol. 277. Berlin: Springer-Verlag, 1987.

Dudowicz, Jacek, Karl F. Freed, and Jack F. Douglas. "The Glass Transition Temperature of Polymer Melts." *The Journal of Physical Chemistry B* 109.45 (2005): 21285-1292.

Earnest, C. M. "Assignment of Glass Transition Temperatures Using Thermomechanical Analysis," *Assignment of the Glass Transition*, ASTM STP 1249, R. J. Seyler, Ed., American Society for Testing and Materials, Philadelphia, 1994, pp. 75-87.

Ediger, M. D. "Spatially Heterogeneous Dynamics in Supercooled Liquids." *Annual Review of Physical Chemistry* 51.1 (2000): 99-128.

Fedors, R.F. "Glass Transition Temperatures and Molecular Weight." *Polymer* 20.9 (1979): 1055-1056.

Fox, T. G., and P. J. Flory. "Second-Order Transition Temperatures and Related Properties of Polystyrene. Influence of Molecular Weight." *Journal of Applied Physics* 21.6 (1950): 581.

Fox, T. G., and S. Loshaek. "Influence of Molecular Weight and Degree of Crosslinking on the Specific Volume and Glass Temperature of Polymers." *Journal of Polymer Science* 15.80 (1955): 371-90.

Fredrickson, Glenn. "Recent Developments in Dynamical Theories of the Liquid-Glass Transition". *Annual Review of Physical Chemistry*, 39, 149 (1988).

Fredrickson, Glenn, and Hans Andersen. "Kinetic Ising Model of the Glass Transition." *Physical Review Letters* 53.13 (1984): 1244-247.

Fulcher, Gordon S. "Analysis of Recent Measurements of the Viscosity Of Glasses." *Journal of the American Ceramic Society* 8.6 (1925): 339-55.

Gerdeen, James C., and Ronald A. L. Rorrer. *Engineering Design with Polymers and Composites*. Boca Raton, FL: CRC, 2012.

Gibbs, Julian H., and Edmund A. Dimarzio. "Nature of the Glass Transition and the Glassy State." *The Journal of Chemical Physics* 28.3 (1958): 373-83.

Goldstein, Martin, and Robert Simha. *The Glass Transition and the Nature of the Glassy State*. Vol. 279. New York: New York Academy of Sciences, 1976.

Götze, W., and L. Sjögren. "α-relaxation near the Liquid-glass Transition." *Journal of Physics C: Solid State Physics* 20.7 (1987): 879-94.

Gupta, P. and Muaro, J. "The Laboratory Glass Transition". *Journal of Chemical Physics*, 126, 22 (2007).

Haynes, William M. *CRC Handbook of Chemistry and Physics: A Ready-reference Book of Chemical and Physical Data*. Boca Raton, FL.: CRC, 2011.

Hiemenz, Paul C., and Timothy Lodge. *Polymer Chemistry*. Boca Raton: CRC, 2007.

Hopkins, Paul, Andrea Fortini, Andrew J. Archer, and Matthias Schmidt. "The Van Hove Distribution Function for Brownian Hard Spheres: Dynamical Test Particle Theory and Computer Simulations for Bulk Dynamics." *The Journal of Chemical Physics* 133.22 (2010).

Huggins, Maurice L. "Solutions of Long Chain Compounds." *The Journal of Chemical Physics* 9.5 (1941): 440.

Katritzky, A.R., S. Sild, V. Lobanov, and M. Karelson. "Quantitative Structure-Property Relationship (QSPR) Correlation of Glass Transition Temperatures of High Molecular Weight Polymers." *Journal of Chemical Information and Modeling* 38.2 (1998): 300-04.

Kauzmann, Walter. "The Nature of the Glassy State and the Behavior of Liquids at Low Temperatures." *Chemical Reviews* 43.2 (1948): 219-56.

Kim, Yong Woo, Jung Tae Park, Joo Hwan Koh, Byoung Ryul Min, and Jong Hak Kim. "Molecular Thermodynamic Model of the Glass Transition Temperature: Dependence on Molecular Weight." *Polymers for Advanced Technologies* 19.8 (2008): 944-46.

Kunal, K., M. Paluch, C. M. Roland, J. E. Puskas, Y. Chen, and A. P. Sokolov. "Polyisobutylene: A Most Unusual Polymer." *Journal of Polymer Science Part B: Polymer Physics* 46.13 (2008): 1390-399.

Le, Tu, V. Chandana Epa, Frank R. Burden, and David A. Winkler. "Quantitative Structure–Property Relationship Modeling of Diverse Materials Properties." *Chemical Reviews* 112.5 (2012): 2889-919.

Lee, Kyung Ju, Do Kyoung Lee, Yong Woo Kim, Woo-Seok Choe, and Jong Hak Kim. "Theoretical Consideration on the Glass Transition Behavior of Polymer Nanocomposites." *Journal of Polymer Science Part B: Polymer Physics* 45.16 (2007): 2232-238.

Leutheusser, E. "Dynamical Model of the Liquid-glass Transition." *Physical Review A* 29.5 (1984): 2765-773.

Lubchenko, Vassiliy, and Peter G. Wolynes. "Theory of Structural Glasses and Supercooled Liquids." *Annual Review of Physical Chemistry* 58.1 (2007): 235-66.

Mark, James E. *Physical Properties of Polymers*. Cambridge: Cambridge UP, 2004.

Más, J., A. Vidaurre, J. M. Meseguer, F. Romero, M. Monleón Pradas, J. L. Gómez Ribelles, M. L. L. Maspoch, O. O. Santana, P. Pagés, and J. Pérez-Folch. "Dynamic Mechanical Properties of Polycarbonate and Acrylonitrile–butadiene–styrene Copolymer Blends." *Journal of Applied Polymer Science* 83.7 (2002): 1507-516.

Meyers, Marc A., and Krishan Kumar Chawla. *Mechanical Behavior of Materials*. Cambridge: Cambridge UP, 2010.

Moynihan, Cornelius T., Allan J. Easteal, James Wilder, and Joseph Tucker. "Dependence of the Glass Transition Temperature on Heating and Cooling Rate." *The Journal of Physical Chemistry* 78.26 (1974): 2673-677.

Ngai, K. L., R. W. Rendell, and D. J. Plazek. "Couplings between the Cooperatively Rearranging Regions of the Adam–Gibbs Theory of Relaxations in Glass-forming Liquids." *The Journal of Chemical Physics* 94.4 (1991): 3018-029.

Ogawa, Toshio. "Effects of Molecular Weight on Mechanical Properties of Polypropylene." *Journal of Applied Polymer Science* 44.10 (1992): 1869-871.

Painter, P. and Coleman, M. *Fundamentals of Polymer Science: an Introductory Text*. Lancaster: Technomic Publishing, 1997.

Palmer, R. G. "Broken Ergodicity." *Advances in Physics* 31.6 (1982): 669-735.

Roland, C. M. "Characteristic Relaxation Times and their Invariance to Thermodynamic Conditions." *Soft Matter* 4.12 (2008): 2316-22.

Scherer, George W. *Relaxation in Glass and Composites*. Malabar, FL: Krieger Pub., 1992.

Seyler, Rickey J. *Assignment of the Glass Transition*. Philadelphia, PA: ASTM, 1994.

Skoog, Douglas A., F. James. Holler, and Timothy A. Nieman. *Principles of Instrumental Analysis*. Philadelphia: Saunders College Pub., 1998.

Taborek, P., R. Kleiman, and D. Bishop. "Power-law Behavior in the Viscosity of Supercooled Liquids." *Physical Review B* 34.3 (1986): 1835-840.

Van Dijk, Menno A., and André Wakker. *Concepts of Polymer Thermodynamics*. Lancaster: Technomic Pub., 1997.

Van Krevelen, D. W., and K. T. Nijenhuis. *Properties of Polymers: Their Correlation with Chemical Structure: Their Numerical Estimation and Prediction from Additive Group Contributions*. Amsterdam: Elsevier, 2009.

Wang, Junmei, and Tingjun Hou. "Application of Molecular Dynamics Simulations in Molecular Property Prediction II: Diffusion Coefficient." *Journal of Computational Chemistry* 32.16 (2011): 3505-519.

Weber, Thomas, Glenn Fredrickson, and Frank Stillinger. "Relaxation Behavior in a Tiling Model for Glasses." *Physical Review B* 34.11 (1986): 7641-651.

Williams, Graham, and David C. Watts. "Non-symmetrical Dielectric Relaxation Behavior Arising from a Simple Empirical Decay Function." *Transactions of the Faraday Society* 66 (1970): 80-85.

Yu, Xinliang, Wenhao Yu, and Xueye Wang. "A Simple Three-descriptor Model for the Prediction of the Glass-transition Temperatures of Vinyl Polymers." *Journal of Applied Polymer Science* 115.6 (2010): 3721-726.

**Appendix A** – Glass and Melting Temperatures for Common Polymers

(Gerdeen et. al, 2012)

| Polymer | Abbreviation | $T_g$ (°C) | $T_m$ (°C) |
|---|---|---|---|
| Polyethylene | PE | -90 to -135 | 115 to 137 |
| Polypropylene | PP | -10 | 176 |
| Polystyrene | PS | 95 | 240 |
| Polyvinyl Chloride | PVC | 85 | 212 |
| Polyvinyl Fluoride | PVF | -20 to 45 | 200 |
| Polyvinylidene Chloride | PVDC | -15 | 198 |
| Polyamide 6 | PA6 | 50 | 215 |
| Polyamide 6 / 6 | PA6/6 | 90 | 260 |
| Poly(methyl methacrylate) | PMMA | 105 | 175 |
| Polycarbonate | PC | 150 | 265 |
| Natural Rubber | NR | -75 | 28 |
| Poly(acrylonitrile-co-butadiene-co-styrene) | ABS | 100 | 230 |
| Polytetrafluoroethylene | PTFE | -65 | 327 |
| Butyl Rubber | BR | -90 | 154 |
| Polyethylene Terephthalate | PET | 68 to 80 | 212 to 265 |

**Appendix B** – Parameter Values and Corresponding T$_g$ Values for Select Polymers Using 5-descriptor QSPR Model (Cao & Lin, 2003 and Katritzky et. al, 1998)

| Polymer | $\Sigma MV_{ter}(R_{ter})$ | $L_F$ | $\Delta X_{SB}$ | $\Delta PEI$ | $Q_{\pm}$ | $T_g(K)_{calc}$ | $T_g(K)_{exp}$ |
|---|---|---|---|---|---|---|---|
| Poly(ethylene) | 0.0 | 0 | 0 | 0 | 0 | 204 | 195 |
| Poly(ethylethylene) | 102.0 | 1 | 0 | 2.2909 | 0 | 256 | 228 |
| Poly(butylethylene) | 102.0 | 3 | 0 | 2.4438 | 0 | 239 | 220 |
| Poly(cyclopentylethylene) | 282.2 | 0 | 0 | 2.7085 | 0 | 339 | 348 |
| Poly(cyclohexylethylene) | 310.5 | 0 | 0 | 2.7683 | 0 | 350 | 363 |
| Poly(acrylic acid) | 139.2 | 0 | 0.3350 | 2.0174 | -0.0483 | 380 | 379 |
| Poly(methyl acrylate) | 102.0 | 2 | 0.2096 | 2.1170 | 0 | 274 | 281 |
| Poly(ethyl acrylate) | 102.0 | 3 | 0.1634 | 2.1703 | 0 | 259 | 251 |
| Poly(sec-butyl acrylate) | 102.0 | 3 | 0.1245 | 2.2567 | 0 | 255 | 253 |
| Poly(vinyl alcohol) | 77.9 | 0 | 0.2737 | 0.8957 | -0.0483 | 337 | 358 |
| Poly(vinyl chloride) | 93.2 | 0 | 0.5126 | 2.1800 | 0 | 331 | 348 |
| Poly(acrylonitrile) | 106.0 | 0 | 0.3209 | 1.9146 | -0.0692 | 388 | 378 |
| Poly(vinyl acetate) | 166.4 | 1 | 0.2096 | 1.2195 | 0 | 300 | 301 |
| Poly(styrene) | 275.2 | 0 | 0.1050 | 2.5699 | 0 | 349 | 373 |
| Poly(2-chlorostyrene) | 307.8 | 0 | 0.1466 | 2.6427 | 0 | 368 | 392 |
| Poly(3-chlorostyrene) | 315.7 | 0 | 0.1466 | 2.6055 | 0 | 371 | 363 |
| Poly(4-chlorostyrene) | 309.0 | 0 | 0.1466 | 2.5908 | 0 | 368 | 389 |
| Poly(2-methylstyrene) | 323.3 | 0 | 0.0958 | 2.6695 | 0 | 368 | 409 |
| Poly(3-methylstyrene) | 327.1 | 0 | 0.0958 | 2.6232 | 0 | 369 | 374 |
| Poly(4-methylstyrene) | 319.6 | 0 | 0.0958 | 2.6030 | 0 | 366 | 374 |
| Poly(4-fluorostyrene) | 290.4 | 0 | 0.1669 | 2.5684 | 0 | 364 | 379 |
| Poly(propylene) | 102.0 | 0 | 0 | 2.0411 | 0 | 262 | 233 |
| Poly(1-pentene) | 102.0 | 2 | 0 | 2.3905 | 0 | 247 | 220 |
| Poly(ethoxyethylene) | 102.0 | 2 | 0.1117 | 1.2451 | 0 | 253 | 254 |
| Poly(tert-butyl acrylate) | 258.1 | 2 | 0.1245 | 2.2769 | 0 | 325 | 315 |
| Poly(n-butyl acrylate) | 102.0 | 5 | 0.1245 | 2.2261 | 0 | 236 | 219 |
| Poly(vinyl hexyl ether) | 102.0 | 6 | 0.0821 | 1.3705 | 0 | 214 | 209 |
| Poly(1,1-dimethylethylene) | 0 | 0 | 0 | 4.0821 | 0 | 241 | 199 |
| Poly(1,1-dichloroethylene) | 0 | 0 | 0 | 4.3600 | 0 | 243 | 256 |
| Poly(1,1-difluoroethylene) | 0 | 0 | 0 | 1.1140 | 0 | 214 | 233 |
| Poly(α-methylstyrene) | 377.2 | 0 | 0.0956 | 4.6110 | 0 | 406 | 409 |
| Poly(methyl methacrylate) | 211.3 | 2 | 0.2084 | 4.1581 | 0 | 335 | 378 |

# Appendix B (continued)

| Polymer | $\Sigma MV_{ter}(R_{ter})$ | $L_F$ | $\Delta X_{SB}$ | $\Delta PEI$ | $Q_\pm$ | $T_g(K)_{calc}$ | $T_g(K)_{exp}$ |
|---|---|---|---|---|---|---|---|
| Poly(ethyl methacrylate) | 211.3 | 3 | 0.1595 | 4.2114 | 0 | 320 | 324 |
| Poly(isopropyl methacrylate) | 258.1 | 2 | 0.1336 | 4.2647 | 0 | 344 | 327 |
| Poly(ethyl chloroacrylate) | 195.2 | 3 | 0.5175 | 4.3503 | 0 | 364 | 366 |
| Poly(2-chloroethyl methacrylate) | 195.2 | 4 | 0.2108 | 4.2323 | 0 | 312 | 365 |
| Poly(tert-butyl methacrylate) | 408.0 | 2 | 0.1174 | 4.3180 | 0 | 401 | 380 |
| Poly(phenyl methacrylate) | 377.2 | 2 | 0.1709 | 4.2929 | 0 | 396 | 393 |
| Poly(chlorotrifluoroethylene) | 162.0 | 0 | 0.4322 | 3.8510 | 0 | 362 | 373 |
| Poly(oxymethylene) | 0 | 0 | 0 | 0 | 0 | 204 | 218 |
| Poly(oxyethylene) | 0 | 0 | 0 | 0 | 0 | 204 | 206 |
| Poly(oxytrimethylene) | 0 | 0 | 0 | 0 | 0 | 204 | 195 |
| Poly(oxytetramethylene) | 0 | 0 | 0 | 0 | 0 | 204 | 190 |
| Poly(ethylene terephthalate) | 58.0 | 0 | 0.7449 | 0.8020 | 0 | 337 | 345 |
| Poly(vinyl n-octyl ether) | 102.0 | 8 | 0.0776 | 1.3915 | 0 | 196 | 194 |
| Poly(vinyl n-decyl ether) | 102.0 | 10 | 0.0748 | 1.4053 | 0 | 178 | 197 |
| Poly(oxyoctamethylene) | 0 | 0 | 0 | 0 | 0 | 204 | 203 |
| Poly(oxyhexamethylene) | 0 | 0 | 0 | 0 | 0 | 204 | 204 |
| Poly(vinyl n-pentyl ether) | 102.0 | 5 | 0.0855 | 1.3541 | 0 | 223 | 207 |
| Poly(vinyl 2-ethylhexyl ether) | 102.0 | 6 | 0.0796 | 1.4238 | 0 | 214 | 207 |
| Poly(n-octyl acrylate) | 102.0 | 9 | 0.0976 | 2.2706 | 0 | 197 | 208 |
| Poly(n-octyl methylacrylate) | 211.3 | 9 | 0.0871 | 4.3117 | 0 | 257 | 253 |
| Poly(n-heptyl acrylate) | 102.0 | 8 | 0.1019 | 2.2634 | 0 | 207 | 213 |
| Poly(n-nonyl acrylate) | 102.0 | 10 | 0.0942 | 2.2773 | 0 | 188 | 216 |
| Poly(n-hexyl acrylate) | 102.0 | 7 | 0.1073 | 2.2550 | 0 | 216 | 216 |
| Poly(1-heptene) | 102.0 | 4 | 0 | 2.4768 | 0 | 230 | 220 |
| Poly(vinyl n-butyl ether) | 102.0 | 4 | 0.0905 | 1.3315 | 0 | 233 | 221 |
| Poly(n-propyl acrylate) | 102.0 | 4 | 0.1393 | 2.2034 | 0 | 247 | 229 |
| Poly(vinylisobutyl ether) | 211.3 | 2 | 0.0905 | 1.3517 | 0 | 293 | 251 |
| Poly(vinyl sec-butyl ether) | 102.0 | 3 | 0.0905 | 1.3980 | 0 | 242 | 253 |
| Poly(pentafluoroethyl ethylene) | 68.8 | 2 | 0.6374 | 2.2442 | 0 | 322 | 314 |
| Poly(2,3,3,3-tetrafluoropropylene) | 137.6 | 1 | 0.6712 | 2.5518 | 0 | 365 | 315 |
| Poly(3,3-dimethylbutylmethacrylate) | 408.0 | 4 | 0.0751 | 4.3124 | 0 | 377 | 318 |
| Poly(n-butyl acrylamide) | 211.3 | 5 | 0.1267 | 4.3411 | -0.0257 | 329 | 319 |
| Poly(vinyl trifluoroacetate) | 68.8 | 3 | 0.5793 | 1.2118 | 0 | 295 | 319 |

# Appendix B (continued)

| Polymer | $\Sigma MV_{ter}(R_{ter})$ | $L_F$ | $\Delta X_{SB}$ | $\Delta PEI$ | $Q_\pm$ | $T_g(K)_{calc}$ | $T_g(K)_{exp}$ |
|---|---|---|---|---|---|---|---|
| Poly(3-methyl-1-butene) | 211.3 | 0 | 0 | 2.5407 | 0 | 309 | 323 |
| Poly($n$-butyl α-chloroacrylate) | 195.2 | 5 | 0.5228 | 4.4061 | 0 | 348 | 330 |
| Poly($sec$-butyl methacrylate) | 211.3 | 4 | 0.1174 | 4.2978 | 0 | 306 | 330 |
| Poly(heptafluoropropyl ethylene) | 68.8 | 3 | 0.6268 | 2.3413 | 0 | 312 | 331 |
| Poly(3-pentyl acrylate) | 102.0 | 4 | 0.1145 | 2.2898 | 0 | 245 | 257 |
| Poly(5-methyl-1-hexene) | 211.3 | 2 | 0 | 2.4971 | 0 | 291 | 259 |
| Poly(oxy-2,2- | 0 | 1 | 0.0083 | 4.5173 | 0 | 237 | 265 |
| Poly($n$-hexyl methacrylate) | 211.3 | 7 | 0.0982 | 4.2960 | 0 | 276 | 268 |
| Poly(vinyl isopropyl ether) | 211.3 | 1 | 0.0981 | 1.3447 | 0 | 303 | 270 |
| Poly[$p$-($n$-butyl)styrene] | 102.0 | 3 | 0.0828 | 2.6546 | 0 | 252 | 279 |
| Poly($n$-butyl methacrylate) | 211.3 | 5 | 0.1174 | 4.2671 | 0 | 297 | 293 |
| Poly(2-methoxyethyl methacrylate) | 211.3 | 5 | 0.1648 | 2.2008 | 0 | 285 | 293 |
| Poly(3,3,3-trifluoropropylene) | 68.8 | 1 | 0.6643 | 1.9948 | 0 | 332 | 300 |
| Poly(4-methyl-1-pentene) | 211.3 | 1 | 0 | 2.4901 | 0 | 300 | 302 |
| Poly(vinyl chloroacetate) | 93.2 | 3 | 0.2831 | 1.2551 | 0 | 264 | 304 |
| Poly($n$-propyl methacrylate) | 211.3 | 4 | 0.1336 | 4.2445 | 0 | 308 | 306 |
| Poly(3-cyclopentyl-1-propene) | 282.2 | 1 | 0 | 2.5469 | 0 | 328 | 333 |
| Poly(3-phenyl-1-propene) | 275.2 | 1 | 0 | 2.5267 | 0 | 325 | 333 |
| Poly($n$-propyl α-chloroacrylate) | 195.2 | 4 | 0.5210 | 4.3834 | 0 | 356 | 344 |
| Poly($sec$-butyl α-chloroacrylate) | 195.2 | 4 | 0.5228 | 4.4367 | 0 | 357 | 347 |
| Poly(3-cyclohexyl-1-propene) | 310.5 | 1 | 0 | 2.5832 | 0 | 340 | 348 |
| Poly(vinyl acetal) | 310.5 | 1 | 0.1810 | 2.0150 | 0 | 360 | 355 |
| Poly(vinyl formal) | 310.5 | 0 | 0.2589 | 1.9154 | 0 | 378 | 378 |

# Appendix C – Molecular Descriptor and Corresponding $T_g$ Values for Select Polymers Using 3-descriptor QSPR Model (Yu, Yu, & Wang, 2009)

| Polymers | $Mv$ | $BIC5$ | $Mor13m$ | $T_g$, (exp) | $T_g$, (calc)[a] | $T_g$, (calc)[b] |
|---|---|---|---|---|---|---|
| poly(acrylic acid) | 0.58 | 0.276 | -0.007 | 379 | 378 | 401 |
| poly(3-thiabutyl acrylate) | 0.58 | 0.852 | -0.063 | 213 | 227 | 244 |
| poly(2-chlorophenyl acrylate) | 0.69 | 0.904 | -0.496 | 326 | 327 | 330 |
| poly(2,4-dichlorophenyl acrylate) | 0.73 | 0.904 | -0.264 | 333 | 340 | 345 |
| poly(2-cyanoisobutyl acrylate) | 0.59 | 0.725 | -0.221 | 324 | 303 | 295 |
| poly(2-cyanoethyl acrylate) | 0.61 | 0.826 | -0.114 | 277 | 273 | 275 |
| poly(5-cyano-3-oxapentyl acrylate) | 0.59 | 0.870 | -0.291 | 250 | 241 | 258 |
| poly(2-cyanoisopropyl acrylate) | 0.60 | 0.659 | -0.222 | 339 | 320 | 320 |
| poly(4-biphenyl acrylate) | 0.68 | 0.807 | -0.877 | 383 | 387 | 371 |
| poly(dodecyl acrylate) | 0.54 | 0.806 | -0.384 | 270 | 259 | 246 |
| poly(2-ethoxyl-carbonyl-phenyl acrylate) | 0.63 | 0.883 | -0.288 | 303 | 282 | 283 |
| poly(2-ethoxyethyl acrylate) | 0.55 | 0.858 | -0.285 | 223 | 233 | 234 |
| poly(ethyl acrylate) | 0.55 | 0.862 | -0.274 | 249 | 231 | 232 |
| poly(4-butoxycarbonylphenyl acrylate) | 0.61 | 0.841 | -0.506 | 286 | 298 | 292 |
| poly(1H,1H-heptafluorobutyl acrylate) | 0.59 | 0.849 | 0.176 | 243 | 227 | 239 |
| poly(2,2,3,3,5,5,5-heptafluoro-4-oxapentyl acrylate) | 0.59 | 0.858 | 0.426 | 218 | 223 | 223 |
| poly(heptyl acrylate) | 0.55 | 0.864 | -0.245 | 213 | 229 | 230 |
| poly(hexadecyl acrylate) | 0.54 | 0.695 | -0.511 | 308 | 311 | 284 |
| poly(hexyl acrylate) | 0.55 | 0.860 | -0.207 | 216 | 227 | 229 |
| poly(isobutyl acrylate) | 0.55 | 0.728 | -0.185 | 249 | 254 | 264 |
| poly(6-cyano-4-thiahexyl acrylate) | 0.60 | 0.873 | -0.162 | 215 | 238 | 258 |
| poly(2-methoxycarbonylphenyl acrylate) | 0.64 | 0.883 | -0.332 | 319 | 299 | 292 |
| poly(4-methoxycarbonylphenyl acrylate) | 0.64 | 0.818 | -0.418 | 340 | 318 | 315 |
| poly(4-methoxyphenyl acrylate) | 0.63 | 0.808 | -0.645 | 324 | 326 | 323 |

# Appendix C (continued)

| Polymers | $M_v$ | BIC5 | Mor13m | $T_g$, (exp) | $T_g$, (calc)[a] | $T_g$, (calc)[b] |
|---|---|---|---|---|---|---|
| poly(*sec*-butyl acrylate) | 0.55 | 0.818 | -0.146 | 250 | 228 | 237 |
| poly(2-methylbutyl acrylate) | 0.55 | 0.848 | -0.200 | 241 | 227 | 232 |
| poly(2-methyl-7-ethyl-4-undecyl acrylate) | 0.54 | 0.833 | -0.406 | 253 | 257 | 240 |
| poly(2-naphthyl acrylate) | 0.68 | 0.885 | -0.752 | 358 | 341 | 342 |
| poly(1H,1H-nonafluoro-4-oxahexyl acrylate) | 0.59 | 0.862 | 0.920 | 224 | 223 | 195 |
| poly(nonyl acrylate) | 0.54 | 0.871 | -0.294 | 215 | 232 | 223 |
| poly(1h,1h,5h-octafluoropentyl acrylate) | 0.59 | 0.880 | 0.094 | 238 | 223 | 235 |
| poly(pentachlorophenyl acrylate) | 0.84 | 0.812 | 0.702 | 420 | 413 | 395 |
| poly(*n*-pentyl acrylate) | 0.55 | 0.855 | -0.193 | 216 | 226 | 229 |
| polyphenylethyl acrylate) | 0.62 | 0.832 | -0.423 | 270 | 303 | 297 |
| poly(phenyl acrylate) | 0.65 | 0.812 | -0.507 | 330 | 325 | 328 |
| poly(tetradecyl acrylate) | 0.54 | 0.748 | -0.443 | 297 | 288 | 266 |
| poly(4,4,5,5-tetrafluoro-3-oxapentyl acrylate) | 0.57 | 0.885 | -0.075 | 251 | 222 | 228 |
| poly(4-tertbutylphenyl acrylate) | 0.61 | 0.688 | -0.655 | 344 | 342 | 343 |
| poly(o-totyl acrylate) | 0.64 | 0.873 | -0.517 | 325 | 311 | 305 |
| poly(2,2,2trifluoroethyl acrylate) | 0.57 | 0.813 | -0.076 | 263 | 231 | 249 |
| poly(3,3,5-trimethylcyclohexyl acrylate) | 0.56 | 0.812 | -0.481 | 288 | 278 | 264 |
| poly(1H,1H-undecafluorohexyl acrylate) | 0.59 | 0.860 | -0.014 | 234 | 228 | 246 |
| poly(5-cyano-3-thiapentyl acrylate) | 0.61 | 0.870 | -0.025 | 223 | 243 | 258 |
| poly(3-chloro-2,2-bis(chloromethyl)propyl | 0.64 | 0.702 | -0.198 | 319 | 325 | 335 |
| poly(4-chlorophenyl acrylate) | 0.69 | 0.812 | -0.403 | 331 | 340 | 350 |
| poly(4-cyanobenyl acrylate) | 0.67 | 0.816 | -0.425 | 317 | 329 | 337 |
| poly(4-cyanobutyl acrylate) | 0.59 | 0.862 | -0.177 | 233 | 235 | 254 |
| poly(4-thiapentyl acrylate) | 0.58 | 0.858 | -0.085 | 208 | 227 | 244 |
| poly(benzyl acrylate) | 0.64 | 0.823 | -0.441 | 279 | 318 | 315 |
| poly(4-cyanophenyl acrylate) | 0.69 | 0.804 | -0.557 | 363 | 354 | 361 |

# Appendix C    (continued)

| Polymers | $M_v$ | BIC5 | Mor13m | $T_g$, (exp) | $T_g$, (calc)[a] | $T_g$, (calc)[b] |
|---|---|---|---|---|---|---|
| poly(3-dimethylaminophenyl acrylate) | 0.61 | 0.806 | -0.831 | 320 | 337 | 320 |
| poly(4-ethoxyl-carbonyl-phenyl acrylate) | 0.63 | 0.827 | -0.460 | 310 | 314 | 308 |
| poly(3-ethoxyl-carbonyl-phenyl acrylate) | 0.63 | 0.883 | -0.504 | 297 | 299 | 294 |
| poly(3-ethoxypropyl acrylate) | 0.55 | 0.862 | -0.274 | 218 | 231 | 232 |
| poly(2-ethylbutyl acrylate) | 0.55 | 0.746 | -0.195 | 223 | 249 | 260 |
| poly(fluoromethyl acrylate) | 0.58 | 0.676 | -0.143 | 288 | 307 | 297 |
| poly(5,5,6,6,7,7,7-heptafluoro-3-oxaheptyl | 0.57 | 0.866 | -0.220 | 228 | 229 | 242 |
| poly(heptafluoro-2-propyl acrylate) | 0.60 | 0.655 | 0.471 | 283 | 315 | 284 |
| poly(2-heptyl acrylate) | 0.55 | 0.859 | -0.193 | 235 | 226 | 228 |
| poly(butyl acrylate) | 0.55 | 0.849 | -0.152 | 219 | 225 | 229 |
| poly(isopropyl acrylate) | 0.56 | 0.655 | -0.127 | 270 | 295 | 288 |
| poly(3-methoxybutyl acrylate) | 0.55 | 0.856 | -0.299 | 217 | 234 | 235 |
| poly(3-methoxycarbonylphenyl acrylate) | 0.64 | 0.883 | -0.431 | 311 | 304 | 297 |
| poly(2-methoxyethyl acrylate) | 0.55 | 0.852 | -0.254 | 223 | 231 | 233 |
| poly(3-methoxypropyl acrylate) | 0.55 | 0.858 | -0.226 | 198 | 228 | 230 |
| poly(methyl acrylate) | 0.57 | 0.583 | -0.068 | 283 | 317 | 312 |
| poly(3-methylbutyl acrylate) | 0.55 | 0.776 | -0.183 | 228 | 238 | 251 |
| poly(2-methylpentyl acrylate) | 0.55 | 0.854 | -0.231 | 235 | 229 | 232 |
| poly(neopentyl acrylate) | 0.55 | 0.625 | -0.175 | 295 | 301 | 292 |
| poly(1H,1H-nonafluoropentyl acrylate) | 0.59 | 0.855 | 0.336 | 236 | 224 | 228 |
| poly(*tert*-butyl acrylate) | 0.55 | 0.536 | -0.317 | 304 | 324 | 325 |
| poly(1H,1H-pentafluoropropyl acrylate) | 0.59 | 0.813 | 0.133 | 247 | 238 | 251 |
| poly(3-pentyl acrylate) | 0.55 | 0.698 | -0.157 | 267 | 265 | 271 |
| poly(2-tertbutylphenyl acrylate) | 0.61 | 0.738 | -0.713 | 345 | 337 | 332 |
| poly(propyl acrylate) | 0.56 | 0.813 | -0.135 | 236 | 230 | 245 |
| poly(7,7,8,8-tetrafluoro-3,6-dioxaoctyl acrylate) | 0.56 | 0.892 | -0.169 | 233 | 223 | 225 |

# Appendix C  (continued)

| Polymers | Mv | BIC5 | Mor13m | $T_g$, (exp) | $T_g$, (calc)[a] | $T_g$, (calc)[b] |
|---|---|---|---|---|---|---|
| poly(5-thiahexyl acrylate) | 0.57 | 0.862 | -0.078 | 203 | 223 | 235 |
| poly(m-totyl acrylate) | 0.64 | 0.873 | -0.530 | 298 | 312 | 306 |
| poly(p-totyl acrylate) | 0.64 | 0.797 | -0.547 | 316 | 326 | 328 |
| poly(5,5,5-trifluoro-3-oxapentyl acrylate) | 0.57 | 0.858 | -0.191 | 235 | 228 | 242 |
| poly(1H,1H-tridecafluoro-4-oxaoctyl acrylate) | 0.59 | 0.870 | 0.083 | 205 | 224 | 238 |
| poly(8-cyano-7-thiaoctyl acrylate) | 0.59 | 0.877 | 0.023 | 214 | 225 | 239 |
| poly(4-thiahexyl acrylate) | 0.57 | 0.862 | -0.043 | 197 | 223 | 233 |
| poly(3-thiapentyl acrylate) | 0.58 | 0.858 | 0.003 | 202 | 224 | 239 |
| poly(vinyl formate) | 0.58 | 0.452 | -0.025 | 304 | 329 | 353 |
| poly(4-cyclohexyl-1butene) | 0.55 | 0.784 | -0.437 | 313 | 277 | 262 |
| poly(vinyl trifluoroacetate) | 0.60 | 0.654 | 0.294 | 319 | 315 | 294 |
| poly(3-methyl-1-butene) | 0.53 | 0.208 | -0.180 | 323 | 361 | 395 |
| poly(3-phenyl-1propene) | 0.63 | 0.799 | -0.440 | 333 | 318 | 314 |
| poly(vinyl n-octyl ether) | 0.53 | 0.865 | -0.301 | 194 | 234 | 218 |
| poly(vinyl n-pentyl ether) | 0.53 | 0.850 | -0.208 | 207 | 227 | 218 |
| poly(vinyl n-hexyl ether) | 0.53 | 0.856 | -0.235 | 209 | 229 | 217 |
| poly(1-hexene) | 0.53 | 0.768 | -0.147 | 223 | 233 | 237 |
| poly(1-heptene) | 0.53 | 0.816 | -0.160 | 220 | 227 | 224 |
| poly(vinyl sec-butyl ether) | 0.53 | 0.775 | -0.328 | 253 | 256 | 245 |
| poly(vinyl ethyl ether) | 0.53 | 0.661 | -0.167 | 254 | 275 | 268 |
| poly(vinyl chloroacetate) | 0.63 | 0.737 | 0.048 | 304 | 317 | 305 |
| poly(5-methyl-hexene) | 0.53 | 0.685 | -0.147 | 259 | 260 | 260 |
| poly(6-methyl-heptene) | 0.53 | 0.745 | -0.198 | 239 | 244 | 246 |
| poly(vinyl isobutyral) | 0.56 | 0.377 | -0.169 | 329 | 342 | 368 |
| poly(vinyl propional) | 0.56 | 0.472 | -0.120 | 345 | 324 | 339 |
| poly(vinyl acetal) | 0.57 | 0.418 | -0.088 | 355 | 333 | 359 |
| poly(vinyl n-butyl ether) | 0.53 | 0.815 | -0.187 | 221 | 229 | 226 |
| poly(vinyl acetate) | 0.57 | 0.654 | -0.055 | 301 | 302 | 292 |
| poly(4-methyl-1-pentene) | 0.53 | 0.528 | -0.128 | 302 | 315 | 303 |

[a] $T_g$ values calculated with the ANN model. [b] $T_g$ values calculated with the MLR model.

# i want morebooks!

Buy your books fast and straightforward online - at one of world's
fastest growing online book stores! Environmentally sound due to
Print-on-Demand technologies.

Buy your books online at

# www.get-morebooks.com

Kaufen Sie Ihre Bücher schnell und unkompliziert online – auf einer
der am schnellsten wachsenden Buchhandelsplattformen weltweit!
Dank Print-On-Demand umwelt- und ressourcenschonend produzi-
ert.

Bücher schneller online kaufen

# www.morebooks.de

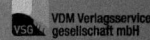
VDM Verlagsservicegesellschaft mbH
Heinrich-Böcking-Str. 6-8     Telefon: +49 681 3720 174     info@vdm-vsg.de
D - 66121 Saarbrücken          Telefax: +49 681 3720 1749    www.vdm-vsg.de

Printed by Books on Demand GmbH, Norderstedt / Germany